TRANSPORT PROCESSES IN BUBBLES, DROPS, AND PARTICLES

TRANSPORT PROCESSES IN BUBBLES, DROPS, AND PARTICLES

Edited by

R. P. Chhabra
Department of Chemical Engineering
Indian Institute of Technology, Kanpur, India

D. De Kee
Department of Chemical Engineering
University of Sherbrooke, Sherbrooke, Que., Canada

●HEMISPHERE PUBLISHING CORPORATION
A member of the Taylor & Francis Group

New York   Washington   Philadelphia   London

TRANSPORT PROCESSES IN BUBBLES, DROPS, AND PARTICLES

1 2 3 4 5 6 7 8 9 0   B R B R   9 8 7 6 5 4 3 2 1

A CIP catalog record for this book is available from the British Library.

Library of Congress Cataloging-in-Publication Data

Transport processes in bubbles, drops, and particles / edited by D. De
Kee, R. P. Chhabra.
     p. cm.
   Includes bibliographical references and index.

    1. Fluid dynamics.   2. Transport theory.   I. De Kee, D. (Daniel)
II. Chhabra, R. P.
QA911.T68   1992
530.4'25—dc20
ISBN 0-89116-999-7

91-23357
CIP

Contents

Preface

This volume brings together a number of authoritative state of the art reviews and contributions, written by well-recognized experts in the field of transport processes, involving bubbles, drops and particles.

Knowledge of the behaviour of bubbles, drops and particles is of vital importance in a variety of chemical and processing applications. Typical examples include flow and mass transfer in systems involving bubbles and drops, atomisation, flow behaviour of particulate suspensions, electrochemical processes, motion of bubbles, drops and rigid particles in rheologically complex media and in low gravity fields with particular reference to space applications, and so on. In addition to the foregoing, the study of the hydrodynamic behaviour of bubbles, drops and particles is challenging and of fundamental importance, as it provides a useful starting point toward the understanding of the more realistic multiparticle systems.

In chapter 1, Subramanian presents an authoritative and critical account of the advances made in the area of bubbles and drops moving in reduced gravity fields. Clearly, the area is still in its infancy. Chapter 2 gives an overview of some of the recent developments on the formation, motion and coalescence of gas bubbles in rheologically complex non-Newtonian continuous phase. Next, Walters and Tanner discuss a classical problem in fluid mechanics, the motion of a sphere in elastic liquids which have so far defied a complete theoretical treatment. This problem—experimentally as well as theoretically—seems to be anything but simple. In chapter 4, Quintana highlights the roles of surfactants on the motion of and mass transfer to/from bubbles and drops. Some ideas on the interactions between surface active agents and the viscoelastic properties of the continuous phase are also presented. Next, Tikuisis and Ward discuss the idea of surface resistance involving mass transfer between a gas and a liquid, and illustrate how it can falsify the measurements of molecular diffusivities based on this technique. Factors influencing the shapes of bubbles and drops in Newtonian fluids are described by Grace and Wairegi in chapter 6. Kaloni and Stastna then deal with the calculation of an effective viscosity of a suspension in a non-Newtonian fluid and with the flow of viscoelastic fluids past fluid spheres. In chapter 8, Carreau discusses the rheology of filled polymeric systems, such as composites and blends. Finally, Ramarao and Tien present a general treatment emphasizing the quantitative evaluation of the deposition of diffusive aerosols from a fluid past a collector. They focus on the determination of particle trajectories and deposition fluxes using the Brownian dynamics simulation method.

With the publication of this work, we hope to update and complement earlier works, as the present volume addresses a diversified range of topics. In particular, electrochemical, space, and non-Newtonian applications are included.

This volume should be of interest to all those engaged in basic as well as in applied research. The information presented herein is equally valuable for practicing engineers, who are constantly dealing with composites and complex situations such as those involving non-Newtonian materials, reduced gravity conditions, and so on. The contents of this volume are accessible to people with a background in engineering and/or in pure sciences.

Finally, we would also like to take this opportunity to thank the contributors who, despite their busy schedules, kindly agreed to participate. We also thank the publishers for their kind and patient assistance.

R. P. Chhabra
D. De Kee

The Editors

R. P. CHHABRA earned the degree of Ph.D. in Chemical Engineering from Monash University, Melbourne, Australia. He is currently Professor of Chemical Engineering at the Indian Institute of Technology, Kanpur, India. Prior to this Dr. R. P. Chhabra taught Chemical Engineering at the University College of Swansea, Great Britain and was a visiting professor at SUNY, Buffalo during 1987.

His current research interests are in the general area of non-Newtonian fluid mechanics, multiphase flows, and transport properties of liquid metals. He has published about 80 technical papers, and in 1988 he won the Indian Institute of Chemical Engineers Amar Dye Chem. Award for excellence in Research. He is a member of numerous professional organizations.

D. DE KEE earned the degree of Ph.D. in Chemical Engineering from the University of Montreal. He is currently a Professor of Chemical Engineering at the University of Sherbrooke. Prior to this, Dr. D. De Kee taught Chemical Engineering in Ontario, and was a Visiting Professor at Columbia University during 1986–87.

His Current research interests are in the general area of rheology. In particular, he is engaged in research concerning the development of constitutive equations, polymer processing, biorheology, and two phase flows. He has published about 90 technical papers and books, and won the Canadian Society of Chemical Engineering ERCO Award in 1987 and the Canadian Society for Chemistry Protective Coatings Award in 1990.

He is a member of several professional organizations and is a fellow of the Chemical Institute of Canada.

Chapter 1

The Motion of Bubbles and Drops in Reduced Gravity

R. SHANKAR SUBRAMANIAN
Department of Chemical Engineering
Clarkson University
Potsdam, New York 13699-5705, USA

1   Introduction

The reader of this book is well aware that bubbles and drops are ubiquitous in their appearance. Thus, it should be no surprise that in scientific experiments as well as in engineering applications in orbiting spacecraft, one encounters them. Orbiting vehicles are in near free fall, and within such a vehicle, gravitational levels are typically three to five orders of magnitude smaller than those encountered on earth [1, 2]. The motion of drops on earth is normally dominated by the action of gravity and the density difference between the drop and the surrounding fluid. This fact has led to a substantial literature devoted to the subject of gravity-driven motion which is discussed in the book by Clift *et al.* [3]. In contrast, with perhaps the exception of electrophoresis of colloidal size bubbles/drops, virtually all of the work on the motion of drops due to mechanisms other than gravity is relatively recent, and the field is in its infancy. The time appears right for providing a brief summary of the state of the literature and pointing out opportunities for new research in this subject. It should be mentioned that the literature on low gravity fluid flows in bulk fluids (in contrast to the motion of suspended objects) has been reviewed by Ostrach [2], and will not be addressed here even though there have been some significant advances made since the appearance of that review. Another resource for the reader is a book by Myshkis *et al.* [4] on the subject of low gravity fluid mechanics. Also, a very helpful survey of drop and bubble motion in reduced gravity by Wozniak *et al.* [5] has appeared recently. Wherever possible, repetition of material covered adequately in that survey will be avoided.

One might wonder why rigid particles have been left out from the present review. Their motion under reduced gravity conditions also is likely to be very important in some applications. However, they lack an important characteristic possessed by bubbles and drops, namely a free interface. The interface between a bubble or a drop and the continuous phase in which it is suspended is usually mobile, permitting tangential motion along the interface. It is this fact that distinguishes such objects from rigid particles in fluid mechanical descriptions. Incidentally, gas bubbles are just a special case of drops wherein the density and viscosity of the interior fluid are small compared to the corresponding properties of the continuous phase. As we shall see later, interface mobility is critical to surface tension driven flow which is perhaps the most important natural mechanism for the motion of drops and bubbles in a reduced gravity environment.

Relatively short durations of low gravity conditions are available in drop towers and drop tubes (2-5 seconds), aircraft flying parabolic trajectories (15-60 seconds), and aboard sounding rockets whose payloads can be recovered (several minutes). However, orbiting vehicles such as the NASA Space Shuttle now provide several days of near free fall time, and one can foresee the availability of many months of free fall for experiments aboard the Space Station, now in the planning stages.

First, I shall try to identify the reason for the current interest in the low gravity fluid mechanics of drops and bubbles. Beginning with the advent of space probes and orbiting spacecraft from about the middle of this century, concern has been expressed about the fate of drops and bubbles either already present in a system or generated due to some physical or chemical process. On earth, when this happens, the density difference and gravity combine to yield a net hydrostatic force on them which causes them to move to the top or bottom of the fluid column in which they are present. In fact, it usually requires considerable effort to keep them from settling. In contrast, orbiting vehicles and space probes are in a near free fall condition. Non-zero residual gravitational levels are still present aboard such vehicles. The reasons are many, and are well discussed in other places [1, 2]. However, these background gravity levels are small, and so one would suppose that any suspended object in a fluid would have a negligible hydrostatic force on it and therefore would virtually stand still. This can be an advantage sometimes, and a problem in other applications.

Consider the attempts to produce a uniform composite solid in space (Gelles and Markworth [6], Ahlborn and Löhberg [7], Gelles et al. [8], Walter [9], Predel et al. [10]). The idea was to start with a binary mixture (for example, Aluminum and Indium) which forms a single liquid phase at a high temperature, but exhibits a miscibility gap at lower temperatures. One would heat the mixture above the consolute temperature so as to form a single liquid phase, and then cool it into the miscibility gap wherein two immiscible liquids are thermodynamically favored. Cooling would therefore lead to the precipitation of a collection of fine drops of a second liquid phase. On earth, these drops would either rise or sink rapidly due to density differences and coalesce resulting in massive phase separation into two layers. It was anticipated that, in reduced gravity, the drops would hardly move. Thus, it was expected that cooling would produce a solid phase with a finely dispersed second solid within it with some novel material properties. In practice, massive phase separation was observed in several low gravity experiments. It was conjectured in [8] that the drops moved and coalesced due to forces other than gravity.

There is considerable interest in crystal growth and solidification of metals in reduced gravity [1]. The principal reason appears to be the anticipated reduction in buoyant convection which is believed to be the source of poor product quality in earth-based processes. However, dissolved gases will be rejected at the melt-solid interface [11] in the form of bubbles. On earth, the buoyant force can be used to move the bubbles away from the interface and prevent their incorporation into the solid. It would be necessary to have alternative means for managing such bubbles in low gravity processing.

2

Another application wherein bubbles are encountered in low gravity experiments is in glass processing. It has been suggested that certain difficult glass-formers can be produced in space because of the ability to process them without a container [12, 13, 14]. However, gas bubbles are evolved copiously when glass is formed due to chemical reactions as well as from the gas trapped among the particles of the raw materials used. These bubbles usually must be removed to produce a useful product. On earth, the bubbles rise to the free surface or dissolve. In space, the former mechanism will be unavailable, and alternative means must be sought to manage gas bubbles [15, 16, 17].

The above examples illustrate scientific experiments in which drops or bubbles might be encountered. There are engineering examples as well. For instance, Ostrach [2] mentions liquid-rockets used to power space vehicles in orbit or space probes. The liquid may evaporate, leading to the formation of vapor bubbles which must be managed. Similarly, the Space Shuttle cooling system uses liquids in which gas bubbles can be nucleated. Suspensions involving immiscible liquids might be involved in a novel heat pipe suggested by Schmid [18].

What forces other than gravity can one imagine that would lead to the motion of suspended objects? If the objects carry a charge, electric fields can produce a force on them. Electrophoresis, which refers to the motion of charged particles in electric fields is a well-studied subject and is considered in detail in books (see, for example, Shaw [19]) as well as in recent review articles [20, 21]. One might envision an analog wherein drops of magnetic fluids (see Rosensweig [22] for a discussion of the fascinating behavior of such fluids) move under the action of a magnetic field even though applications for this are likely to be less common. Nothing more will be said about either topic in the present review.

2 Migration Induced by Interfacial Forces

The interface between two fluids appears to be in a state of tension. A detailed discussion of the origin of interfacial tension from a molecular point of view may be found elsewhere (see the book by Miller and Neogi [23] and the references cited therein). For our purposes, it is sufficient here that we regard the interface as a geometric surface of zero thickness separating two phases. In particular, I might note that the interfacial tension at a fluid-fluid interface can depend on temperature and composition. Therefore, any variation in either will lead to a gradient in interfacial tension. As a result, the interface can exert tangential forces on the fluids on either side leading, in general, to motion of both fluids.

Bulk fluid motion driven by the interface is a topic which has been studied for over a century. Scriven and Sternling [24] provide a historical review of the extensive literature on this subject up to 1959. While these authors discuss interfacial phenomena in the context of the motion of drops and bubbles, they confine their attention to the retardation of the motion of such objects in a gravitational field. This was a topic which attracted considerable attention in the earlier part of the present century. Small drops and bubbles were observed to move at lower speeds than those predicted by theory. The most commonly accepted explanation is that due to Frumkin and

Levich [25] discussed in detail by Levich [26]. The physical picture involves the adsorption of surface active chemicals on the interface, which are then swept to the rear of the drop as it moves through the surrounding fluid. A dynamic balance is established between the transport of surfactant by convection and diffusion along the interface and its transport to the interface from the bulk fluids. Consequently, a gradient of surfactant concentration is established along the interface, and the resulting interfacial tension gradient opposes the motion of the fluid at the interface. In some cases, an insoluble surfactant can form a stagnant cap on the rear surface of a drop translating through a fluid [27, 28]. In all cases, the drop exhibits more resistance to motion than one would expect if the interface were completely mobile. The original analyses have been refined in recent years. The reader is referred to a detailed review given by Harper [29], recent articles by Holbrook and LeVan [30, 31, 32], Chang and Berg [33], and Sadhal and Johnson [34], and the review by Quintana [35] appearing in this volume for additional information. Incidentally, another explanation advanced for the anomalous motion of drops is based on temperature variations which arise on a drop surface because of creation of fresh surface in the forward half and destruction of surface in the rear half. The hypothesis is that the resulting surface tension variations oppose motion along the interface in the same way that surface tension gradients from surfactant concentration gradients would. However, the temperature variations involved were shown to be too small by Harper et al. [36] who investigated this hypothesis in detail.

In addition to the review given by Scriven and Sternling [24], I might mention that another general review of interfacial tension driven phenomena has been written by Levich and Krylov [37] and one pertaining to flows driven by the interface in the context of crystal growth by Schwabe [38].

2.1 Gas Bubble Experiments

2.1.1 Experiments on earth

In 1959, Young, Goldstein, and Block reported on the results of some interesting experiments on gas bubbles [39]. This article laid the foundation for virtually all of the subsequent work to be described in the rest of the present section. Therefore, it is worthwhile to describe the contents of [39] in some detail. Young et al. were motivated by the recognition that when a gas bubble is placed in a liquid possessing a temperature gradient, the interfacial tension at the bubble surface will vary with position due to its dependence on temperature. Typically, the surface tension is at a minimum at the warm pole and a maximum at the cold pole. The resulting gradient in surface tension leads to a tangential stress at the interface in the direction of the cold pole. The neighboring fluid is dragged in this direction by the stress. As a consequence, the bubble propels itself in the opposite direction, namely, that of the temperature gradient. Therefore, bubbles should experience an attraction toward hot regions in fluids, and their behavior under gravity should be moderated by this phenomenon. Since the interfacial tension depends on species concentration at the interface, one can expect similar behavior when concentration gradients are induced at a bubble surface by maintaining a gradient of species concentration in the surrounding fluid. However, temperature gradients are much simpler to control and maintain, and therefore all of the research to date in this area has involved the use of temperature gradients rather than composition gradients. Some interesting experiments on the spreading of surfactant on a drop surface due to surface tension gradients are reported by Chifu et al. [40]. These authors comment that work is in progress on the motion of drops driven by these surfactant composition gradients. Composition gradients can give rise to an additional contribution to the velocity of the drop due to diffusiophoresis [41, 21].

Young et al. set out to demonstrate the effect of a vertical temperature gradient on bubbles by confining a small column of liquid between the anvils of a machinist's micrometer. Several Dow-Corning DC-200 series silicone oils were used for the experiments, and some control experiments were performed in n-hexadecane. The anvils were surrounded by copper blocks, and the lower block could be heated by passing a current through nichrome wire wrapped around it. The temperature gradients were kept sufficiently small to avoid the Rayleigh instability from the unfavorable density stratification in the liquid. Thermometers inserted into the copper blocks were used to record the temperatures of the upper and lower anvils. Air bubbles were introduced into the liquid column and were found to collect at the lower anvil. This abnormal behavior qualitatively confirmed the authors' expectations regarding the action of surface tension gradients. They slowly reduced the vertical temperature gradient until a relatively large bubble from the lower anvil was seen to detach and rise slowly. Another adjustment of the temperature gradient permitted the authors to hold the bubble nearly motionless midway between the anvils. The bubble diameter was measured with a traveling microscope. The authors thus measured the temperature gradient necessary to hold bubbles of various diameters from rising under earth's gravity. In control experiments with n-hexadecane, a trace amount of silicone oil was added and found to "prevent completely the abnormal bubble behavior" thus leading the authors to conclude that the effect observed by them was indeed a surface tension effect. The presumption was that the contaminant adsorbed on the bubble surface and formed an insoluble monolayer which supported the tangential stress from the temperature gradient. We'll postpone a discussion of the role of surfactants to a later point.

Young et al. opened up a substantial research area which thrives at the present time stimulated by applications in reduced gravity. As mentioned earlier, a steady temperature gradient is relatively straightforward to induce and maintain in a fluid in contrast to a steady concentration gradient. This has led to a number of experiments that have already been conducted on ground, and several planned for space flight in the next decade. The experiments of Young et al. are remarkable in their simplicity and yet produced results of substantial impact. For instance, the authors verified that the temperature gradient required to hold a bubble stationary increased with the bubble size and was independent of viscosity as they predicted from a theoretical model which they also presented in their article. While their data show considerable scatter due to reasons which are well-discussed in subsequent articles by others, the general trends were unmistakable. I should add that vapor bubble motion driven by a temperature gradient is well-illustrated along with several other surface tension driven fluid mechanical phenomena in a color film by L. Trefethen [42].

While Young et al. chose to work with gas bubbles, the phenomenon they demonstrated is equally applicable to liquid drops. Motion driven by interfacial tension gradients is usually referred to as "capillarity driven motion" or simply "capillary motion." When the gradients are the result of temperature variations, we use the term "thermocapillary motion." When composition variations cause the gradients, the rather awkward expression "diffusocapillary motion" has been used. Interestingly, even though one can predict that drops and bubbles will move due to composition gradient effects, no clear demonstration of this phenomenon has yet been made due to the difficulty in producing and maintaining such gradients.

Young *et al.* [39] also developed a theoretical description of the problem. They considered the motion of a fluid sphere under the combined action of gravity and a **downward** temperature gradient. Under conditions of negligible convective transport of momentum and energy (limit of zero Reynolds and Peclet numbers), they solved the Stokes and Laplace equations for the velocity and temperature fields in and around the drop respectively, and specialized the solution using the boundary conditions for the problem. From a balance of the forces on the drop, they obtained a result for the migration velocity of the drop which, after correcting some minor typographical errors, is given below.

$$ v = \frac{2}{3\,\mu\,(2+3\alpha)} \left[\frac{3}{2+\beta} \frac{|\,\nabla T_\infty\,|\;\sigma_T}{} R \; + \; (\rho - \rho')\,g\,(1+\alpha)\,R^2 \right]. \tag{1} $$

Here, ρ is the density of the continuous phase fluid, ρ', the density of the drop phase fluid, μ, the dynamic viscosity of the continuous phase fluid, g, the magnitude of the acceleration due to gravity, R, the drop radius, σ_T, the rate of change of interfacial tension with temperature, and $|\,\nabla T_\infty\,|$, the magnitude of the downward temperature gradient imposed in the continuous phase fluid. α stands for the ratio of the viscosity of the drop phase to that of the continuous phase, and β represents a similar ratio of thermal conductivities. A negative value of v in Equation (1) implies **downward** motion.

For a gas bubble, the viscosity and thermal conductivity of the gas are both quite negligible compared to the same properties of the surrounding liquid. Thus, one might set the property ratios, α and β to zero in the above result. Also, one might consider the density of the gas negligible compared to that of the continuous phase. It is worth noting that Young *et al.* believed that their result was only good when the velocity of the bubble is set equal to zero (see the footnote on page 355 of their article). In actuality, their solution is the first step in a perturbation scheme for small values of the Reynolds and Peclet numbers, and should be a good approximation when the values of these parameters are small. Note that when the velocity in the left hand side of Equation (1) is set equal to zero, the temperature gradient needed to hold the bubble still is linear in bubble radius and independent of the viscosity of the continuous phase. Despite the scatter in the data of Young *et al.*, this trend is clearly evident.

The experiments of Young *et al.* had certain pitfalls. Presumably, the experiments were conducted with sufficiently small bubbles that the liquid could be considered infinite in extent (so that the results could be compared with their theory which assumed this), but bubble-bubble interactions could not always have been avoided.

More importantly, bulk circulation in the liquid column would have been inevitable. There are two reasons for this. First, there was a temperature gradient along the free surface of the liquid column which would have caused a gradient of interfacial tension leading to upward flow at this surface and a downward return flow in the core. Also, heat loss to the surrounding air would have resulted in a warmer core leading to buoyant upward motion in the core and a downward flow at the free surface. Depending upon the relative magnitude of each of these opposing influences, there must have been some residual motion along the axis of the column where Young *et al.* made their measurements. Also, the observations were made through a curved fluid surface, and the authors do not state that any corrections were applied for optical distortions. Finally, temperature measurement and control might not have been sufficiently precise. All of these factors could have contributed to the scatter observed in the results plotted by Young *et al.*. One might note that the authors were aware of some of these complications since they state that "measurements were confined to the region near the vertical axis of the liquid sample in order to minimize any effects associated with the free cylindrical surface."

Kuznetsov *et al.* [43] presented a theoretical analysis similar to that of Young *et al.* [39]. However, they omitted an analysis of the energy equation, stating that "during the motion of the bubble, the temperature gradient inside it keeps a constant value." They reported the migration speed in terms of this gradient in the absence of gravity. They also performed qualitative experiments using an air bubble in water, notorious for surface contamination, and convinced themselves that they did indeed observe the predicted motion. No comparison with theory was attempted. Since the water was contained in a horizontal tube with a horizontal temperature gradient applied to it, one might conjecture that free convection might have been the principal, if not the only, source of the motion observed by these authors. Interestingly, the authors apparently were unaware of the work of Young *et al.* until close to the time of publication of their article.

Hardy [44] built an experimental apparatus eliminating the conditions which probably caused much of the scatter in the data of Young *et al.*. His experiments were performed in a closed rectangular cell which was completely filled with the test fluid, a Dow-Corning silicone oil. He introduced one bubble at a time and made observations. Presumably due to variations in the solubility of air in the silicone oil, Hardy found the bubbles grew and shrank alternately as they migrated in his cell. In fact, a typical bubble was sufficiently large when it was introduced that it rose to the top and came to rest against the upper surface of the cell. The liquid there was relatively cool and presumably undersaturated, leading to shrinkage of the bubble. The bubble shrank in size until it detached from the top surface and started moving downward while continuing to shrink for a while. Then the bubble started growing (when it reached supersaturated liquid) while still moving downward. However, the bubble finally became sufficiently large that it slowed and stopped, and then reversed its direction of motion. Hardy reported that these position and size oscillations would continue with regularity.

The basic reason for the above observations is that for motion with negligible inertial effects, the buoyant velocity of a bubble is proportional to the square of its radius whereas the thermocapillary velocity (which is in the downward direction in Hardy's experiments) is proportional to the first power of the radius. Therefore, at a certain size the bubble would have zero velocity. At larger sizes, under otherwise identical conditions, its motion would be dominated by gravity, and at smaller sizes, by thermocapillarity. Therefore, the bubble rested against the top surface until it shrank to a size where the thermocapillary force became sufficiently large relative to the buoyant force to start moving the bubble downward. Similar arguments apply near the bottom of the cell where the bubble grew to a sufficiently large size for buoyancy to stop the downward motion and reverse its direction.

Hardy concentrated his measurements on conditions where the bubble had negligible velocity, that is when it reversed direction. Under these conditions, he measured the bubble diameter and the gradient needed to hold the bubble stationary. He found the latter to depend linearly on the former as expected from Equation (1) if one sets the left hand side to zero. By fitting the data to a straight line, Hardy inferred a value of the gradient of surface tension with temperature. He made independent measurements of the surface tension of his test liquid as a function of temperature using the pendant drop method and found the surface tension gradient with temperature obtained from the two techniques to be in reasonable agreement. Hardy also reported some data on bubble migration velocities, but found them to be somewhat smaller than those predicted by theory.

Recently, R. Merritt, working on his doctoral thesis research in my laboratory, measured the migration velocities of air bubbles in Dow-Corning DC-200 series silicone oils in an apparatus fashioned after Hardy's cell [45]. Bubbles ranging in diameter from 100 to 200 μm were subjected to a downward temperature gradient ranging up to 9 K/mm in three different silicone oils. The bubble trajectories were recorded on videotape using a microscope with a video camera attachment. The images were later analyzed on a frame-by-frame basis using electronic cross-hairs. From the videotapes, bubble velocities were measured as a function of bubble diameter for each temperature gradient in each silicone oil. To test the theoretical predictions of Young et al. [39] regarding the dependence of the bubble velocity on bubble size, the results were plotted so as to yield a straight line, the slope of which was the surface tension gradient. The experiments are described by Merritt and Subramanian in [46]; they report good agreement with the theory since the experiments were indeed conducted at very small values of the Reynolds and Peclet numbers. They too observed the bubbles to change in size during the course of an experiment. Typically, a bubble would double in size over a period of a minute and a half. In spite of this, the results were in agreement with the theory which assumed a bubble of constant size.

McGrew et al. [47] measured the force exerted on a bubble attached to a cantilever wire and held fixed in a liquid with a temperature gradient. They found that the thermocapillary force (obtained by subtracting the buoyant force from that measured by the apparatus) scaled consistently with the temperature gradient in n-butanol, but was larger than that predicted by theory in methanol. They attribute the discrepancy in methanol to the high vapor pressure in this case. However, more studies are needed to resolve the question.

Wilcox et al. [48] discuss screening experiments for thermocapillary bubble motion in liquid subjected to a temperature gradient in horizontal tubes. The main constraint here was to assure that the motion of the bubbles occurs much more rapidly than the background convection which will be present in such a geometry. The authors report interesting observations on experiments performed in various organic liquids. In some cases, the bubbles were found to move in the "wrong" direction which the authors suggest is either because of a positive coefficient of surface tension with temperature or preferential evaporation of a species. Results from a similar set of experiments performed on glass melts contained in a rectangular channel are given by Mattox *et al.* [49]. These authors suggest that in glass sealing and enamelling operations, gas bubbles are often removed independent of orientation of the melt. They indicate that this might be because of thermocapillary migration of bubbles to the hot melt surface. However, caution must be exercised in accepting this explanation since free liquid surfaces in the presence of lateral temperature gradients can give rise to bulk circulation flows as well.

2.1.2 Experiments in Low Gravity

Very few low gravity experiments on the motion of gas bubbles have been conducted to date. Some early experiments were performed aboard NASA rocket flights by Wilcox and co-workers [50, 51]. These authors applied a temperature gradient to a sodium borate melt contained in a long narrow cell of rectangular cross-section and photographed the motion of small gas bubbles suspended in the melt through a microscope. The objective of the experiments was to demonstrate that the bubbles would move in the glass melt due to the temperature gradient. The bubbles were indeed observed to move toward warmer regions, and larger bubbles moved more rapidly than smaller bubbles. Also, bubbles accelerated as they moved into warmer melt due to the reduced viscosity. Precise quantitative comparisons were not attempted due to experimental complications such as interactions among bubbles and with neighboring surfaces. The above observations were in contrast to those of Papazian *et al.* [11, 52] who also conducted low gravity experiments aboard earlier rocket flights on bubbles in carbon tetrabromide. In that case, bubbles were observed not to move in a temperature gradient, and the authors advanced an explanation for their observations based on contamination by surface active agents.

We now have clear evidence of surface tension driven motion of bubbles in various liquids based on ground-based experiments. Therefore, the reason for performing low gravity experiments would be to go beyond the range of parameters attainable on ground. Non-dimensionalization of the governing equations for bubble motion in a temperature gradient reveals that the scaled bubble speed depends on two key parameters – the Reynolds number and the Peclet number, known as the Marangoni number when motion is caused purely by thermocapillary means. The former represents the relative importance of convective transport of momentum compared to molecular transport, and the latter measures the relative importance of convective transport of energy compared to molecular transport. In practical situations in space applications, it is expected that a wide range of values of each of these groups will be encountered depending on the physical properties of the fluids involved. However, virtually all of the experimental work on ground has been directed at small or negligible values of these two groups. To understand the reasons, let us examine the definitions.

$$Re = \frac{R \, v_o}{\nu} , \tag{2}$$

$$Ma = \frac{R \, v_o}{\kappa} . \tag{3}$$

Here, ν is the kinematic viscosity of the continuous phase, and κ is its thermal diffusivity. R is the radius of the bubble. In the theory of Young et al. [39], for purely thermocapillary migration, the velocity of the bubble is one half of the characteristic velocity, v_o, which is defined below.

$$v_o = \frac{|\sigma_T| \, |\nabla T_\infty| \, R}{\mu} . \tag{4}$$

The problem with using experiments on ground under conditions of large Peclet number is that they will not necessarily be representative of experiments in reduced gravity under conditions of large Marangoni number. There are fundamental differences in the types of flow fields obtained when motion is driven by a body force so that a non-zero hydrodynamic force is necessary to achieve steady migration, as opposed to situations when the motion is driven by the interface under conditions of zero hydrodynamic force. This, in turn, will affect the temperature fields in the surrounding fluid, and therefore the driving force for thermocapillary motion which is the temperature field on the drop surface. Therefore, experiments on ground, even if they can be conveniently performed, are of limited utility in predicting the behavior of drops and bubbles in reduced gravity. However, there are difficulties in performing even these experiments on ground which I shall briefly mention below.

Achieving relatively large values of the Peclet number requires that the diameter and the velocity of the bubble must both be relatively large. Ground-based experiments must use a vertical temperature gradient. If there are lateral components of the temperature gradient, they will always give rise to buoyant convection in the liquid which is usually quite complex, and cannot simply be subtracted out from the observed bubble motion. Vertical temperature gradients can be downward or upward. Only a downward temperature gradient will oppose the natural buoyant rise of a bubble so that it can be observed for a sufficiently long period of time. Upward temperature gradients have not been used in the past since bubbles of the sizes needed would rise too rapidly even in the absence of such a gradient. A temperature gradient which contributes to upward motion will only make the rise more rapid, making observation and measurement more difficult. Finally, even when a vertical temperature gradient is used, sidewall heat loss will result in some lateral temperature gradients, and it is not possible to completely eliminate buoyant convection. Thus, in ground-based experiments, it is important to make measurements to be certain that the observed bubble velocities are much larger than the background convection levels in the cell.

Downward temperature gradients unfortunately give rise to unstable density stratification of the continuous phase. That is, the fluid is more dense at the top than at the bottom. In such systems, if the Rayleigh number exceeds a certain critical value, there would be complex convection patterns (see Chandrasekhar [53]) which will make it impossible to infer the bubble speed in a quiescent fluid. The Rayleigh number depends on the fourth power of the depth of the liquid and thus one is normally restricted to a depth of a few millimeters [46] in common liquids. This severely restricts the largest size of bubble one can use while still avoiding interactions with the top and bottom cell surfaces, and keeps the achievable Reynolds and Peclet numbers to values on the order of unity or smaller.

Thompson [54] was motivated by the above fact to design and perform experiments in the drop tower at NASA Lewis Research Center (LeRC) on nitrogen bubbles injected into ethylene glycol, a Dow-Corning silicone oil, ethanol, and water, and subjected to a temperature gradient. The results are reported in Thompson et al. [55]. The free fall experiments lasted approximately five seconds. The bubbles were between 6 and 8 mm in diameter, and the container was a Lexan cylinder 120 mm tall and of diameter 120 mm. Prior to each free fall experiment, the test liquid in the container was heated from above for a period ranging from 210 to 285 minutes. From a sample temperature profile given in [55], it appears that the temperature field in the system was approaching linearity in the upper half of the container. After the start of free fall, the bubbles were injected from the bottom of the container, at the center, and appear to have had a significant residual velocity when released. Thompson also reports that the bubbles oscillated when introduced, but that the oscillations decayed rapidly, especially in the more viscous liquids. Quite often, two bubbles were released unintentionally; data from these runs also are reported. Several isothermal runs were made for comparison purposes. Bubbles in these runs were found to move throughout the run, but slowed as the run progressed. In all cases, the bubbles were observed to remain spherical after the decay of the oscillations.

Based on his observations, Thompson [54] concluded that the bubbles in water did not move appreciably due to the action of the temperature gradient. However, the temperature gradient had a significant effect on bubble motion in the other three liquids. For comparison with his experimental data, Thompson used the prediction of Young et al. [39] as well as a result he obtained using perturbation theory for including small convective transport effects. Unfortunately, most of his experiments were performed under conditions wherein such effects must have been large.

In the limit of zero gravity, Equation (1) reduces to the following result. For convenience, I have introduced a minus sign to make the result positive for common systems wherein the surface tension decreases with increasing temperature.

$$ v = \frac{2 \mid \nabla T_\infty \mid (-\sigma_T) R}{\mu (2 + 3\alpha)(2 + \beta)} . \tag{5} $$

From the theory of Young et al. one would expect Equation (5) to hold only for negligible values of both the Reynolds and Marangoni numbers. However, Crespo and Manuel [56], and independently Balasubramaniam and Chai [57], showed that this prediction for the bubble velocity is good for all values of the Reynolds number as long as the Marangoni number is negligible. From Thompson's data on ethylene glycol, one finds that up to Reynolds number values of 5.66 and Marangoni number values as large as 713, the observed velocities were consistent with Equation (5). This is very puzzling. However, the data from experiments in silicone oil show velocities

11

smaller than those predicted from Equation (5) for a Reynolds number of 3.9 and a corresponding Marangoni number of 534. Smaller velocities than those predicted by Equation (5) also are observed for bubbles in ethanol wherein the Reynolds number was as large as 111.6 with a corresponding Marangoni number of 1,674. Thompson also included data from multiple bubble runs where the leading bubble appeared to be unaffected by the trailing bubble. If one considers these runs, the maximum value of the Reynolds number reached (with ethanol) was 570, and the corresponding Marangoni number was 8,550. Incidentally, Thompson defined the Reynolds number as the Marangoni number, and the scaled speed of the bubble as the Reynolds number. In the above, I have used the definitions given in Equations (2,3). For further details, and a useful discussion of the past literature, the reader is referred to Thompson's thesis [54].

Siekmann and co-workers have reported on experiments on bubble motion in a temperature gradient conducted on the D-1 Shuttle mission [58, 59, 60]. The authors conducted experiments on air bubbles as well as water drops injected into silicone oil contained in two different cells. It appears that the water drops did not move. A collection of bubbles was introduced into the silicone oil after setting up a temperature gradient, and the authors indicate that temperature profiles were estimated from a numerical solution of the one-dimensional energy equation.

In [58], a table of measured velocities is reported, and these are compared with values predicted from Equation (5). From this table, one observes that experimentally measured velocities were substantially larger than the predicted values for Marangoni number values of the order of unity. However, at a relatively large value of the Marangoni number (288), the measured velocity was approximately one-tenth the predicted value. As mentioned earlier, the predictions are from a theory good only for negligible values of the Marangoni number. The authors report additional information in the paper indicating that the velocities of two bubbles experiencing the same temperature gradient were roughly proportional to their radii. In [59], the authors show a graph of observed bubble speeds, scaled using v_o, plotted against the Marangoni number. They compare the data against new theoretical predictions obtained by solving the governing equations numerically, and the graph indicates good agreement. In [60], very similar information is given, but including uncertainty estimates on the experimental data points. Comparison of the information in the graph in [59, 60] against the table in [58] indicates that the data in the table and the data in the graph may have come from different sets of bubbles. The Prandtl number of the silicone oil used is reported to be 687 which implies that in the runs, inertial effects were relatively unimportant. This is confirmed by the values of the Reynolds number reported in the table in [58] which lie in the range 0.0013 to 0.42. The authors report that in all cases, the bubbles remained spherical.

2.2   Liquid Drop Experiments

Interest in liquid drops moving due to interfacial forces appears to have been stimulated directly by applications in space processing. B. Facemire appears to have performed the first experiments on this subject, and the results were reported in a NASA Technical Memorandum by Lacy et al. [61]. The principal objective of this research was to study the phase separation behavior in binary systems exhibiting a miscibility gap. By using transparent model fluids, optical observations could be made. The system studied was a mixture of diethylene glycol (DEG) and ethyl salicylate (ES). This mixture forms a homogeneous phase at high temperatures, but when cooled below 70°C, drops of ES-rich liquid appear. Normally, such drops are more dense than the surrounding DEG-rich liquid, and settle out. Lacy *et al.* subjected the ES-rich drops to an upward temperature gradient, and observed that they migrated upward. The resulting speeds were measured and compared with predictions from the theory of Young *et al.* [39]. The drops were found to move less rapidly than predicted by theory. Even though a comparison with theory was presented, the experimental conditions did not permit proper testing of the theory.

K. D. Barton, another doctoral student working in my laboratory, has made measurements of drop migration speeds in this system using single drops of pure ES injected into pure DEG in a test cell similar to that of Merritt. Barton triply distilled the fluids to avoid contamination problems, and used video techniques for recording and analyzing his data. Since ES drops are slightly more dense than DEG, the drops should normally sink when inserted into the DEG, and they do under isothermal conditions. However, in an upward temperature gradient (which permits stable stratification), the drops were found to migrate upward. Addition of small amounts of Triton X-100, a common surfactant, to both the ES and the DEG resulted in virtually complete elimination of the effect, confirming that the observed motion is driven by the interface. The conditions of the experiments were such that convective transport effects could be considered negligible. The results, reported in Barton and Subramanian [62] show agreement with the predictions from Equation (1) in the correct scaling of the thermocapillary migration velocities with drop radius and temperature gradient.

Interestingly, Barton and Subramanian [62] found that the temperature coefficient of the interfacial tension inferred from their experiments differed significantly from that reported by Lacy *et al.* [61] who used a DuNouy ring method. They suggest that the discrepancy could be due to trace contaminants adsorbed on the interface. The role of insoluble surfactants in influencing the thermocapillary motion of droplets has been analyzed by Kim and Subramanian [63, 64]. In part I, these authors consider the case where convective transport of surfactant on the interface is very rapid. In this limit, a stagnant surfactant cap can form on the drop. This leads to reduction of the thermocapillary migration velocity due to two reasons. First, only the mobile part of the interface can lead to thermocapillary motion of the drop. Secondly, the drop offers more resistance to motion of the continuous phase fluid past it because of the presence of the stagnant cap. However, as long as the cap does not cover the entire drop surface, thermocapillary motion can still result. In part II, Kim and Subramanian consider the more general case where surfactant can be transported by both convection and diffusion along the interface, and obtain numerical solutions of the governing surface transport equation for the surfactant. The interested reader should consult these references as well as a more recent article by Nadim and Borhan [65] for further details. I should mention that the surfactant contamination argument has often been used to explain experimental observations wherein bubbles or drops either do not move at all in a temperature gradient, or move too slowly [11, 56].

Activity on liquid drops which began in the early eighties in Europe, still continues. The work is well-discussed in the review article by Wozniak *et al.* [5] and will only be briefly mentioned here. Langbein and Heide [66] performed experiments in a binary liquid mixture containing cyclohexane and methanol. This system exhibits a miscibility gap, and the experiments, performed on low gravity rocket flights, showed drops moving in the direction of the temperature gradient. No quantitative comparison with theory was attempted. Delitzsch et al. [67] as well as Wozniak [68] have reported results on some interesting experiments. The idea here was to choose two liquids carefully so that their densities are exactly matched at a certain temperature. The drop phase is chosen such that it is more dense than the continuous phase at higher temperatures. Then, in an upward temperature gradient, the drop will move up to where the downward net gravitational force on it exactly matches the upward thermocapillary force and remain there. Examination of Equation (1) reveals that smaller drops would come to rest farther upward in a given temperature gradient. Delitzsch et al. verified this using drops of water in both butylbenzoate and fluorobenzene under conditions where Equation (1) should be valid. They inferred the surface tension gradient in each case from their experiments. The experiments of Wozniak [68] were conducted on large drops of paraffin oil (2 to 8 mm in diameter) in aqueous ethanol solutions. This resulted in conditions such that Equation (1) would not be applicable. In very recent experiments, Hähnel et al. [69] have extended the earlier work of Delitzsch et al. on water drops in butyl benzoate. The authors found systematic deviations from the predictions of Equation (1) for the equilibrium location of the drops. They surmise that the deviations are due to the importance of convective energy transport in their experiments which is ignored in the theory of Young et al. [39].

No theoretical results are currently available for comparison with experiments on liquid drops under the simultaneous action of a gravitational field and a temperature gradient when convective transport becomes important. Finally, it is interesting to note that water was used in [67, 68, 69] which showed positive results in spite of the fact that water is notorious for contamination by trace amounts of surface active chemicals [29].

Very recently, Rashidnia and Balasubramaniam [70] have reported results on the thermocapillary migration of vegetable oil drops in silicone oil when subjected to an upward temperature gradient. This is another system in which densities are matched at a certain temperature, and above this temperature, the drop phase is more dense than the continuous phase. The drops ranged in diameter from 3 to 10 mm, and were found to move upward toward warmer regions as expected. They came to rest at an appropriate location where the forces were in balance. Rashidnia and Balasubramaniam observed persistent circulation within the drops and monitored it using tracers. They found the direction of flow to be consistent with the sign of σ_T determined from independent measurements of the interfacial tension as a function of temperature. At long times, they observed the drops to move downward toward colder regions, presumably due to mass transfer effects which changed the average density of the drops. While the observed equilibrium drop locations were in qualitative agreement with the predictions of Young et al. [39], systematic quantitative deviations were noted. Also, the authors report interfacial velocities which are typically smaller by a factor of 2 or 3 when compared with the theoretical predictions. They suggest several reasons for the discrepancies including significant convective energy transport effects, interactions with neighboring walls, buoyant convection in the vicinity of the drops, and possible consequences from mass transfer. Finally, Rashidnia and Balasubramaniam also report that in similar experiments using water-methanol mixtures for the drop phase, flow within the drops was considerably retarded when compared to that in the vegetable oil drops even though vegetable oil is much more viscous than the methanol-water mixture.

All of the above work was motivated by applications to materials processing in space. Interestingly, Maris et al. [71] were studying the nucleation rate for solidification of supercooled liquid hydrogen held in pressurized ^{4}He fluid. They produced drops of liquid hydrogen which reached equilibrium locations in the continuous phase, and the position changed with size in a manner consistent with Equation (1). The experiments are similar to those of Wozniak [68] in that the Reynolds numbers were much larger than unity, and the drops reached stationary locations. One would expect that the Marangoni numbers were large as well, even though the authors do not actually give values of this group. These authors also found interesting drop interactions about which they make qualitative comments.

The only low gravity thermocapillary migration experiments on liquid drops reported to date were carried out by Wozniak and Siekmann [72, 73] in two separate experiments in the European sounding rocket (MASER) program. The authors report a summary in [72], and give more details in [73]. The continuous phase and the drop phase were exactly the same ones used earlier by Wozniak [68]. A table given in [73] shows that data were obtained on a total of four drops ranging in diameter from 1.63 to 3.1 mm in the two experiments. In each case, preheating was begun two minutes before liftoff of the rocket, and heating was continued throughout the flight which provided 7 minutes and 10 seconds of low gravity conditions. After approximately five minutes into the low gravity period, which the authors indicate was sufficient to establish steady temperature profiles in the continuous phase, the drops were injected, and their trajectories followed using video techniques.

In the first rocket experiment (MASER1), two drops of paraffin oil were injected, but the first drop did not move at all while the second one moved in the right direction. The authors suggest that the interface of the first drop was probably contaminated because of extended contact between the drop and the continuous phase. In the second rocket experiment (MASER2), several gas bubbles were accidentally introduced into the cell and moved toward the hot wall and remained there throughout the experiments. The first attempt to inject a drop resulted instead in the introduction of a bubble which was found to migrate toward the hot wall, where it remained. Subsequently, three drops were introduced. The Reynolds number in the two rocket experiments ranged from 1.9 to 8 and the Marangoni number varied from 43 to 183. Interestingly, the experimentally measured velocity is close to the prediction from Equation (5) for the drop with the largest set of values of the Reynolds and Marangoni numbers while more deviations are seen in a case with intermediate values of these parameters. Given the nature of the discrepancies between theory and experiment in the ground-based experiments to date on liquid drops, one should not attempt to draw significant conclusions from this observation regarding the sufficiency of the theoretical model in correctly accounting for the phenomena involved.

From a personal communication from G. Wozniak, I have recently learnt that a third rocket experiment (MASER3) was performed in April 1989, with results to be published in the near future. The author indicates that "Nine paraffin droplets of increasing diameter moved toward the heating side of the aqueous ethanol matrix with speeds roughly proportional to the droplet radius. However the droplet speeds are remarkably lower than theory predicts."

Since the Reynolds and Peclet numbers are based on characteristic velocities and not on the drop migration velocity, it is possible to achieve large values of these groups with experiments on **stationary** drops of the type performed in density-matched systems on earth. However, there are difficulties in obtaining theoretical descriptions of these systems. Furthermore, for space applications, the stationary drop experiments are not directly useful. One needs to perform migration experiments over a wide range of values of the Reynolds and Marangoni numbers similar to those of Wozniak and Siekmann [73]. It is worth noting that in **all** of the experiments reported to date, whether in low gravity or on earth, no measurable deformation of shape from a sphere has been observed.

2.3 Theoretical Work on Single Drops/Bubbles

I have already made reference to some theoretical analyses. In this section, I shall discuss a few more articles to put the problems in perspective and describe some interesting features of the theoretical problems. In general, predictions in this field are far ahead of experimental work for reasons explained earlier. Opportunities for low gravity experiments have been limited.

The analysis of the thermocapillary migration problem requires the writing of the governing equations of conservation of mass, momentum, and energy, along with the associated boundary conditions. In general, Newtonian flow with constant density and viscosity is assumed. Viscous dissipation, being unimportant in these problems, is always neglected. Further, the relevant thermal properties, namely the thermal conductivity and the specific heat, are assumed constant as well. Out of necessity, the surface tension is assumed to vary with temperature; otherwise, there would be no driving force for drop motion! However, even here, the usual procedure has been to assume a constant value for the gradient of surface tension with temperature. For gas-liquid interfaces this is not unreasonable provided the variation of temperature over the bubble surface is not larger than, say, $20°C$. In contrast, in liquid-liquid systems, the surface tension sometimes varies nonlinearly with temperature, and if the temperature variation over the drop surface is more than a few $°C$, this assumption may be questionable.

Commonly, a uniform temperature gradient field (analogous to a uniform gravity field in settling problems) is assumed to exist in the undisturbed continuous phase fluid. Thus, the temperature varies with position. Furthermore, the drop continually moves into regions of changing temperature, and therefore, physical properties. Thus, one might wonder about the suitability of assuming constant physical properties. The usual presumption is that the motion occurs sufficiently slowly that the velocity and temperature gradient fields around the drop and within it have reached their "steady" distributions for the prevailing physical property conditions in the vicinity of the drop. Therefore, one might expect minimal error in using some average values of the properties existing in the vicinity of the drop provided the variation of temperature is not too steep. For instance, Merritt and Subramanian [46] as well as Barton and Subramanian [62] used the temperature in the undisturbed fluid in the equatorial plane of the bubble/drop in evaluating the physical properties used in Equation (1). This was found to be satisfactory in comparing theory with experiment.

The basic quasi-steady assumption discussed above was used by everyone who obtained theoretical predictions. However, recently, Dill and Balasubramaniam [74] have analyzed the evolution of the speed of a gas bubble from zero to the quasi-steady value given in Equation (1), while assuming constant physical properties. Their efforts have provided an understanding of the time scales for relaxation of the velocity and temperature gradient fields to their asymptotic distributions in the surrounding liquid while still retaining sufficient simplicity for a primarily analytical technique to be employed.

The assumption of constant physical properties permits one to pose the fluid mechanical problem in as simple a form as possible, permitting the temperature field to influence the flow only via the temperature gradient on the interface. This temperature gradient results in a gradient of interfacial tension which enters the jump momentum balance at the drop surface. In contrast, the velocity field appears in the convective transport terms in the energy equation. The general problem is non-linear, and the momentum and energy equations must be solved together, along with the associated boundary conditions. In problems including gravitational effects, the hydrodynamic force on the drop is calculated and balanced against the hydrostatic force to obtain the quasi-steady migration speed. Most analyses assume the drop to be spherical, and small deformations from that shape have been calculated via perturbation methods.

The reader may consult papers by Bratukhin [75], Thompson et al. [55], Subramanian [76, 77], and Balasubramaniam and Lavery [78] for the governing equations and boundary conditions as well as their scaling. When the equations are scaled, the convective transport terms in the momentum equation are multiplied by the Reynolds number, and similar terms in the energy equation are multiplied by the Peclet number. In problems wherein motion is driven only by thermocapillarity, the Peclet number is the Marangoni number. When the speed of the drop is significantly influenced by gravity as well, one has to be careful in choosing a reference velocity and v_o defined in Equation (4) is not always suitable for normalizing velocities.

It is convenient to work in a reference frame attached to the moving drop. In this case, the temperature of the fluid far away from the drop changes with time even though its temperature gradient is steady. To accommodate this, a scaled temperature is defined using the difference of the actual temperature from that existing at infinity at the drop equator and normalizing via the temperature change at infinity over a distance equal to the drop radius. Bratukhin appears to have been the first to perform the scaling carefully, and discover that this introduces a source term (multiplied by the Marangoni number) in the scaled equations. My first article on the gas bubble problem [76] will provide the interested reader with more detail.

The analysis of Young et al. [39] was analogous to that performed independently by Hadamard and Rybczynski earlier this century for the problem of the steady motion of a drop driven by gravity. The Hadamard-Rybczynski analysis is discussed in the book by Happel and Brenner [79]. In addition to the uniform gravitational field in their analysis, Young *et al.* assumed a uniform temperature gradient field in the undisturbed fluid, aligned parallel to the gravity vector and pointing in the same direction. They completely neglected all convective transport effects. This corresponds to setting the Reynolds and Peclet numbers equal to zero in the scaled equations. While they assumed a spherical shape, they appear to have verified that this was indeed correct since they include the balance of normal stress at the bubble surface as one of their boundary conditions. This is analogous to the result obtained by Taylor and Acrivos [80] in the body force driven motion problem where they find that when inertia is negligible, the spherical shape is consistent with the balance of normal stresses at the drop surface.

Young *et al.* imply through a footnote in their paper that they believed their theoretical predictions to be good only when the bubble is stationary. However, it is easily seen that their result is the leading order term in a perturbation series wherein convective effects are treated as being small. In fact, even when a bubble or drop is held stationary due to the opposing actions of gravity and a temperature gradient, there will be motion in the surrounding fluid. If the Reynolds and/or Peclet numbers defined using a suitable characteristic velocity are large, one cannot justify ignoring the convective transport terms. Thus, the applicability of the result of Young *et al.* does not depend upon whether the drop moves, but rather on whether one can neglect convective transport effects compared to those of molecular transport.

There are some fundamental difficulties in extending the analysis of Young *et al.* to problems including convective transport while retaining gravitational effects. These are discussed in detail by Merritt [45] who tried to analyze the problem in the limit of negligible Reynolds number, but finite Peclet number, by both numerical and asymptotic methods. In the following, I shall confine attention to extensions in which only purely thermocapillary motion is considered.

Interestingly, when one neglects gravity, the irrotational velocity field solution of Young *et al.* for flow in the continuous phase is an exact solution of the full Navier-Stokes equations for any value of the Reynolds number, and not necessarily for small values. This appears to have been noted first by Crespo and Manuel [56]. The requirement is that the surface tension distribution which drives the flow should be axisymmetric and proportional to $\cos\theta$ where θ is the polar angle, measured from the direction of drop motion, in spherical polar coordinates with origin at the drop center. In thermocapillary motion driven by a uniform temperature gradient field, this leads to the conclusion that the temperature field should satisfy Laplace's equation, so that the Marangoni number must be negligible, a fact pointed out by Crespo and Manuel [56]. Balasubramaniam and Chai [57], who made the same discovery independently, discuss this matter in further detail, and use the solution to calculate small deformations of the shape of the drop from that of a sphere. Balasubramaniam and Chai find that a gas bubble undergoing thermocapillary migration should be deformed into an oblate spheroid whereas a drop more dense than the surrounding fluid should become prolate. The analysis is restricted to sufficiently small values of the Prandtl number so that the Marangoni number (which is the product of the Prandtl number and the Reynolds number here) can remain small even though the Reynolds number is substantial. The deformation calculations are limited to small values of the Capillary (Ca) and Weber (Wb) numbers defined below. It is worth noting that $Wb = Ca \times Re$.

$$Ca = \frac{\mu \, v_o}{\sigma}, \tag{6}$$

$$Wb = \frac{\rho \, v_o^2 \, R}{\sigma}. \tag{7}$$

Incidentally, the scaled speed of a bubble also will depend on one of these groups in addition to depending on the Reynolds and Marangoni numbers.

In 1975, Bratukhin [75] attempted to extend the analysis of Young et al. [39] by a straightforward perturbation technique. That is, after scaling the governing equations, he assumed the Prandtl number to be fixed, and wrote formal power series expansions in the Reynolds Number for the field variables. He also permitted small deformations of shape from a sphere. Interestingly, Bratukhin does not cite the work of Young et al. Instead, he refers to an article by Levich and Kuznetsov [81] in which the authors predict the velocity of a drop in a liquid due to a gradient in surfactant concentration. The analysis by Levich and Kuznetsov parallels others given in the book by Levich[26] wherein various problems in which surfactants influence drop motion are analyzed. Basically, under appropriate assumptions, the surfactant concentration on the drop surface is proportional to $\cos \theta$ where θ is the polar angle defined earlier. The problem then becomes analogous to that analyzed by Young et al. who solve Laplace's equation to obtain a temperature field on the drop surface which also is proportional to $\cos \theta$.

Bratukhin carried his perturbation calculations to $O(Re)$ and concluded that at this order, the correction to the drop velocity is zero. He also drew inferences about the shape of the drop. Interestingly, even though the analysis of Bratukhin produces a velocity field good to $O(Re)$ for fixed Prandtl number, and that of Balasubramaniam and Chai leads to a velocity field good for all values of Re but at zero Prandtl number, both yield qualitatively the same prediction regarding the nature of the shape change.

Thompson et al. [55] extended Bratukhin's analysis to $O(Re^2)$. However, at this level, problems appear. In Thompson's thesis [54], one finds a correction to the temperature field at $O(Re^2)$ which does not fall off to zero at infinity. Since the scaled temperature field becomes unbounded linearly with distance from the drop as one approaches infinity (except on the equator), this is mathematically acceptable. However, the situation will become worse at higher orders, and is a symptom of a deeper problem in the analysis. A straightforward perturbation scheme assumes that terms multiplied by the small parameter can be ignored everywhere in the domain. However, in certain problems with unbounded domains, there are regions where the neglected terms become comparable to the included terms. Thus, the expansions in the perturbation parameter are not uniformly valid everywhere. A common method that is used in this class of problems is that of matched asymptotic expansions.

When I first became acquainted with the thermocapillary migration problem, I was interested in viscous fluids with high Prandtl numbers. Therefore, I proceeded to try to extend the analysis of Young et al. [39] assuming negligible Reynolds number, but

19

finite values of the Marangoni number. The results of this work, carried out using the method of matched asymptotic expansions, are given in [76] for a gas bubble. Also, in [76], a brief review of the literature to that point on the gas bubble problem is provided, and comments are made regarding the pitfalls of the straightforward perturbation method. The bubble was assumed spherical in shape. The principal result from reference [76] is one for the velocity of the bubble. When this velocity is scaled using the reference velocity v_o, the resulting scaled velocity of the bubble may be written as follows:

$$ U = \frac{1}{2} \left[1 - \frac{301}{7,200} Ma^2 \right]. \tag{8} $$

In [77], I extended the above analysis to liquid drops. The result is very similar in nature. Again, the drop velocity is scaled by v_o.

$$ U = \frac{2}{(2 + 3\alpha)(2 + \beta)} \left(1 - Z(\alpha, \beta, \gamma) Ma^2 \right), \tag{9} $$

where

$$ Z(\alpha, \beta, \gamma) = \frac{1}{22,050 \, (1 + \alpha)(2 + 3\alpha)^2 (3 + 2\beta)(2 + \beta)^3} $$
$$ \times \, [7 \, (12,642 + 3,619\beta - 3,070\beta^2) + 7\alpha \, (13,818 + 4,431\beta - 3,210\beta^2) $$
$$ - \, 2\gamma \, (11,916 + 17,925\beta + 6,510\beta^2) - 2 \, \alpha \, \gamma \, (10,404 + 16,665\beta + 6,510\beta^2)]. \tag{10} $$

In the above, $\gamma = \beta/\lambda$ where λ is the ratio of the thermal diffusivity of the droplet to that of the continuous phase. Therefore, γ represents the ratio of the heat capacity per unit volume of the droplet phase to that of the continuous phase. Note that the correction to the result of Young et al. [39] in the above results appears at $O(Ma^2)$. It is negative in the case of a gas bubble, but can be negative or positive for a drop depending upon the values of the physical property ratios. Merritt carried the perturbation analysis for the gas bubble to $O(Ma^4)$ and reported the results in an Appendix in his thesis [45]. Nobody has attempted to continue the analysis for liquid drops to higher orders to this date. Shankar and Subramanian [82] reconsidered the problem of the thermocapillary migration of a gas bubble at values of the Marangoni number ranging from 0 to 200, but still at negligible values of the Reynolds number. They solved the energy equation numerically using the method of finite differences. Therefore, they were able to establish the limits of utility of the perturbation result I had obtained earlier which is given in Equation (8). They found that it was only useful for values of the Marangoni number on the order unity or less. In fact, they found that Merritt's result, available at that time to $O(Ma^3)$, still was not useful for $Ma > 2$. However, by applying series improvement techniques to the $O(Ma^3)$ series, Shankar and Subramanian obtained the following result which was indistinguishable from the numerical result for $Ma \leq 20$.

$$U = \frac{1}{2} - 0.1125 \, M^2 - 0.1123 \, M^3. \tag{11}$$

Here,

$$M = \frac{Ma}{(Ma + 2.32)}. \tag{12}$$

Also, for values of $Ma \geq 20$, Shankar and Subramanian found an empirical fit of their numerical results.

$$U = \frac{1.59}{(Ln \, (Ma) + 1.84)}. \tag{13}$$

The results given in Equations (11,13) have not been tested against experiments yet. The extraordinary simplicity of Equation (13) suggests a form for the velocity in this limit which can be used to advantage in constructing asymptotic expansions good for large values of the Marangoni number, a task that has not been accomplished to date.

One should note that Equations (11,13) predict a decrease of the scaled bubble velocity when the Marangoni number is increased. Since the Marangoni number is proportional to the square of the bubble radius, one might be tempted to conclude that increasing the radius will decrease the bubble speed; quite the contrary. The bubble velocity in the above equations is scaled using the reference velocity v_o which depends linearly on the bubble radius. A little algebra will show that the actual bubble velocity will increase with increasing bubble radius, but not linearly.

Attempts to determine an asymptotically valid result for the bubble velocity for large values of the Reynolds and Marangoni number have been made. Crespo and Manuel [56] used dissipation arguments to obtain a result which is two-thirds the value given by Equation (5) with α and β set equal to zero. However, a key step in arriving at their result is the estimation of the surface tension gradient distribution at the bubble surface. This is obtained using arguments which are flawed. A deeper analysis is given by Balasubramaniam [83] who used similar energy arguments, but correctly rejected the approach of [56] for estimating the surface tension gradient. Balasubramaniam [83] first made a careful order estimate and a subsequent analysis to conclude that in the limit of large Reynolds number, the work done by the surface tension forces has to be dissipated predominantly outside the momentum boundary layer. It is worth noting that in this problem with a mobile boundary, at large values of the Reynolds number, the velocity field from potential flow is good everywhere as a leading order approximation; the momentum boundary layer near the bubble surface is necessary only to accommodate the stress discontinuity arising from the neglect of viscous effects. As a consequence the correction to the potential flow velocity field in the boundary layer is a small one.

Balasubramaniam [83], used an integral approach to tackle the energy equation and made some subsequent approximations. He obtained a result for the asymptotic bubble velocity at large values of Reynolds and Marangoni numbers which has the same form as that of Crespo and Manuel [56], but has a smaller numerical coefficient. However, this result needs independent verification in view of the approximations used. Furthermore, certain subtle features of this problem are not included in the treatment. At large values of the Marangoni number, the thermal boundary layer plays the same role that is played by the momentum boundary layer in the large Reynolds number situation. Thus, the temperature change in the boundary layer is small, and as the boundary layer coordinate approaches infinity, the temperature field in the boundary layer must approach the field at the bubble surface in an outer solution wherein conduction is completely neglected. This outer solution has not been obtained. Furthermore, modifications are necessary in the vicinity of the stagnation points wherein boundary layers will exist in which conduction is important, and across which the temperature changes need not be small.

Finally, it is noteworthy that Equation (13), obtained from limited numerical results, predicts that the asymptotic scaled velocity approaches zero as the Marangoni number approaches infinity for the case when the Reynolds number is negligible. However, the results of [56, 83] both are non-zero in the limit as the Marangoni number approaches infinity and the Reynolds number approaches infinity. The issue is unresolved at this stage due to a lack of asymptotic analytical solutions in this limit. Even the numerical calculations discussed below are not of help in settling the question. They indicate that the scaled velocity decreases with increasing values of the Marangoni number at fixed Reynolds number values, be they small or large. On the other hand, they also show the scaled velocity increasing with increasing Reynolds number at fixed values of the Marangoni number.

As mentioned above, some numerical solutions of the governing equations have been reported in the case when convective transport of momentum and energy are both important. Szymczyk and Siekmann [84] have made detailed calculations solving the Navier-Stokes and energy equations numerically by the method of finite differences. Results were obtained for a maximum value of $Ma = 1,000$ and $Re = 100$. These numerical results have been extended by Balasubramaniam and Lavery [78] who carried their calculations up to $Ma = 1,000$ and $Re = 2,000$. These authors confirm the accuracy of the earlier calculations of Szymczyk and Siekmann even though the references cited precede [84]. Both groups find that the scaled bubble velocity is more sensitive to the Marangoni number at fixed Reynolds number than to Reynolds number at fixed Marangoni number. At fixed Prandtl number, the scaled velocity decreases with increasing Reynolds number, and hence with increasing Marangoni number. If one fixes the Reynolds number and investigates the behavior of the scaled velocity, it is found to decrease substantially with increasing Marangoni number. However, as mentioned before, at fixed Marangoni number, the scaled velocity increases with increasing Reynolds number at large values of Reynolds number. It appears that similar numerical results are currently being calculated for the case of liquid drops. Sample results are given by Wozniak et al. [85].

Since large values of the Reynolds number might involve turbulent flow around the drop/bubble, and since the numerical investigations did not include the possibility of turbulence, one should be cautious in using the large Reynolds number results from the above investigations. Also, typically drops and bubbles in common fluids deform substantially when moving at large Reynolds number because the Weber number becomes sufficiently large. All of the numerical calculations to date have assumed the bubble/drop to be a sphere.

So far, I have considered the motion of a drop placed in a uniform temperature gradient field. It would seem that a more complicated field in the undisturbed fluid would be impossible to handle by analytical means. In [86], I have shown how Equation (1) still applies in this case provided one uses the undisturbed temperature gradient evaluated at the drop location, as long as convective transport of energy and momentum are still negligible. Reference [86] also gives some general results for problems in which an arbitrary but known surface tension gradient exists on a drop surface.

To summarize, considerable advances in predicting the thermocapillary velocity of an isolated drop in an unbounded fluid have been made since the appearance of the paper of Young et al.. A large range of values of the Reynolds and Marangoni numbers has been investigated numerically, in addition to analytical explorations by perturbation techniques. Some analyses have taken into account small deformations of the drop shape from that of a sphere. When gravity is included, the only extension of the theory of Young et al. [39] to date appears to be that attempted by Merritt [45]. There are no theoretical predictions currently available with which the experimental results of Wozniak [68] and Rashidnia and Balasubramaniam [70] can be compared.

I would like to conclude this section by commenting on a somewhat unique feature of the flow patterns in the case of a bubble or drop undergoing thermocapillary migration in an unbounded fluid at negligible Reynolds number. This was discovered purely by accident by Shankar and Subramanian [82] who chose to plot representative streamlines. Prior plots of streamlines in a reference frame attached to the bubble showed no interesting qualitative features. When they tried plotting in the laboratory reference frame, they found the patterns shown in Figure 1 which is reproduced here from [82] with permission from Academic Press. The streamlines are shown in a meridian section.

The values of the Stokes streamfunction, ψ, reported in the figure are scaled using $v_0 R^2$ as the reference. For a definition of the Stokes streamfunction and its properties, the reader may consult [79]. For our purposes, it is sufficient to know that the velocity vector is tangent to the streamlines everywhere. Note the perfect fore-aft symmetry in the plot for $Ma = 0$. The key feature I wish to point out is the appearance of a region of reverse flow in the wake of the migrating bubble when the Marangoni number is non-zero. This clearly has important implications for interactions among bubbles, for a trailing bubble caught in this region might move away from the leading bubble, whereas if it were close enough, it might have an opportunity to approach the leading bubble.

As explained by Shankar and Subramanian [82], any arbitrary axisymmetric temperature field on the bubble surface can be written in an infinite series of Legendre Polynomials which form a complete orthogonal basis set. Only the weighting factors multiplying each Legendre Polynomial will change from one problem to another. Therefore, even though the actual temperature field on the bubble surface and elsewhere is determined by the solution of the energy equation which involves a knowledge of the velocity field, we can draw some important conclusions by studying the flow patterns generated by the pure modes, that is, the individual Legendre Polynomials

$P_n(s)$, where $s = cos\theta$. This is possible because the fluid mechanical problem is linear. The gradient of the surface temperature field leads to a surface tension gradient distribution on the bubble surface which appears as a tangential stress applied to the liquid surrounding the bubble. When the surface temperature field is described by a pure Legendre mode, the angular dependence of the resulting streamfunction turns out to be a pure Gegenbauer polynomial. The Gegenbauer Polynomials are integrals of the Legendre Polynomials, and their properties are given in [79]. The necessary mathematical results are given in [76] where I provided a solution for the streamfunction in the thermocapillary migration problem in the form of an infinite series. Each term in that series is the streamfunction for flow driven by a pure Legendre mode of the surface temperature distribution with a multiplicative constant.

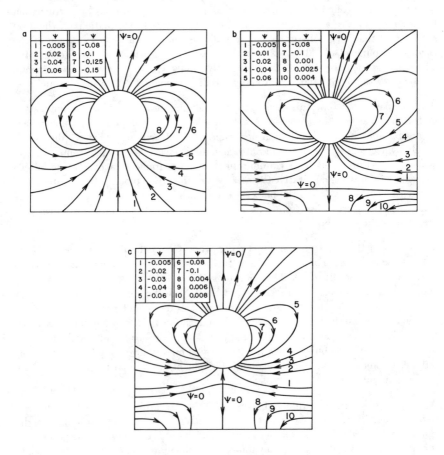

FIGURE 1. Representative Streamlines for $Re = 0$. (a)$Ma = 0$ (b) $Ma = 5$ (c) $Ma = 100$.

Streamlines for the first few modes in a meridian section are given in Figure 2 below. For interpreting the streamfunction values, the reader should note that the streamfunction is scaled in the same way as in Figure 1, and the driving force is a scaled surface temperature equal to $P_n(s)$. The temperature scaling was given earlier in this section.

A key feature to be noted is that the velocity field produced by a surface temperature field proportional to the Legendre Polynomial $P_1(s)$ decays as $1/r^3$ where r is distance measured from the center of the bubble. The velocity field produced by all the higher $P_n(s)$ decay as $1/r^n$. Therefore, the velocity field resulting from the $P_2(s)$ mode of the surface temperature decays slowest, as $1/r^2$. The next more rapid decay will be

FIGURE 2. Representative Streamlines for flow driven by scaled temperature field $P_n(\cos\theta)$.   (a) $n = 1$ (b) $n = 2$ (c) $n = 3$ (d) $n = 4$.

that of the field from the P_1 and P_3 modes, the next will be that from the P_4 mode, and so on. This means that if one goes sufficiently far from the bubble, the flow structure will be dominated by the flow pattern from the P_2 mode. This is true no matter how small the weighting assigned to this mode relative to the P_1 or the higher P_n modes because one can always find a sufficiently large value of r.

Now, examine Figure 2. The flow driven by the P_1 mode has perfect fore-aft symmetry. The bubble pushes fluid out of the way in the direction in which it moves, and there is a return flow behind the bubble. This is the flow pattern from the solution of Young et al. [39] wherein the surface temperature only has a P_1 component.

Introduction of convective energy transport excites higher modes of the surface temperature. In principle, there are contributions from all the modes $P_n(s)$ for all values of n. However, far from the bubble, the dominant contribution to the flow is from the P_2 mode. The figure shows flow toward the bubble in the polar regions, and flow away from it at the equator. The actual direction of this flow will be reversed if the contribution from the P_2 mode is multiplied by a negative constant. This is, in fact, the case in the problem considered by Shankar and Subramanian [82]. Therefore, in their problem, the flow from the P_2 mode actually involves fluid being brought in from the equatorial regions at infinity and pumped out in the polar regions.

Now, imagine a surface temperature dominated mostly by the P_1 mode, but with a small, negative, P_2 contribution. Far from the bubble, in front of the bubble, the flows from the two modes will reinforce each other. However, in the rear, the two flows oppose each other in the polar region. Therefore, at some distance away from the bubble, the flow from the P_2 mode will take over and cause reverse flow. This leads to a dividing streamline separating the two regions of flow, and we get a flow pattern such as that shown in Figure 1b. In Figure 1c, note that the dividing streamline has moved closer to the bubble when compared with Figure 1b, presumably because of an increasingly important role played by the P_2 mode; this occurs because convective energy transport becomes more important due to the increase in the Marangoni number relative to Figure 1b.

I have focused here on the case of negligible Reynolds number when the momentum equations are linear because the superposition arguments require such linearity. However, it should be noted that when the Reynolds number is not small, as in the case considered by Balasubramaniam and Lavery [78], reverse flow regions are observed as well.

It is only incidental that the object in the example chosen above is a gas bubble. Since the discussion is centered on the flow in the continuous phase, the arguments would apply equally well for the case of liquid drops. The importance of the observation in the preceding paragraph becomes clear when one recognizes that higher modes of the surface temperature field can be excited due to a variety of reasons. The solution of Young et al. [39] is a very special case possessing sufficient symmetry to yield a surface temperature which is a pure P_1 mode. Examples of the formation of dividing streamlines and reverse flow regions may be found in cases where a drop is near a surface as in Meyyappan [87] and Barton and Subramanian [88], or when a drop contains an eccentrically located droplet [89]. In these problems, convective effects are negligible, but the higher modes of the surface temperature are excited by the lack of the symmetry which is present in the problem considered by Young *et al.*. The comments made above apply equally well in cases wherein motion is driven by surface tension gradients caused by variations in species concentrations.

2.4   Interactions with Neighboring Surfaces or Objects

In practical applications, one might suspect that bubbles might interact with nearby bubbles and/or surfaces. This motivated a series of theoretical analyses in the same limit considered by Young *et al.* [39], namely, that of negligible convective transport effects. Here, I shall briefly discuss this literature and mention some recent experimental work in this area as well.

Interactions of bubbles, drops, and particles with each other and neighboring boundaries is a subject with a rich literature. Happel and Brenner [79] provide a good introduction to this topic. For motion with negligible inertial effects wherein the governing equations are linear (and are called Stokes equations), superposition can be used in constructing solutions, and therefore, substantial effort has been devoted to this limit.

For the axisymmetric motion of two spheres in an unbounded fluid, an exact solution of Stokes equation for the streamfunction field can be written in bipolar coordinates. This was done by Stimson and Jeffery [90] in 1926. Their solution has formed the basis for a large number of articles wherein the constants were specialized to different sets of boundary conditions to account for rigid and free surfaces. A sphere moving normal to a plane surface is a special case wherein one of the spheres has an infinite radius, and can be handled using the same general solution. The most general cases of two unequal sized drops of different fluids in a third fluid were solved by Rushton and Davies [91, 92] and independently by Haber *et al.* [93]. Brenner [94] solved for the motion of a rigid sphere normal to a rigid plane surface, and Bart [95] extended this solution to fluid drops and plane fluid-fluid interfaces. O'Neill [96] provided a significant extension of the work of Stimson and Jeffery to the non-axisymmetric situation, and his work has spawned several publications extending it in various ways. In all the above cases, the situation involved body-force driven motion in an isothermal fluid.

One of my doctoral students, M. Meyyappan, solved various problems involving the interactions of gas bubbles with each other and with neighboring plane surfaces for bubbles undergoing thermocapillary migration in the limit of negligible convective transport of momentum and energy. Meyyappan *et al.* analyzed the case of a gas bubble moving normal to a plane solid or free liquid surface in [97] and that of two gas bubbles moving along their line of centers in [98]. Both are axisymmetric problems and were solved using bipolar coordinates. Thus, Meyyappan *et al.* specialized the constants in the Stimson-Jeffery solution of the Stokes equation. In thermocapillary migration, since the temperature gradient along the bubble surface leads to the interfacial tension gradient which appears as a discontinuity in the balance of tangential stress across the interface, it is necessary to solve the energy equation as well. For negligible convective transport, the energy equation reduces to the Laplace equation. A general solution in bipolar coordinates was presented by Jeffery [99], and Meyyappan *et al.* specialized this as needed. The solutions were quasi-static in the sense that Meyyappan *et al.*, like others, calculated **steady** fields of velocity and temperature assuming relaxation to such steady states to occur very rapidly when compared to the time scale for the geometric configurations to change.

Later, Meyyappan and Subramanian solved two non-axisymmetric problems. In [100] they analyzed the motion of two bubbles oriented arbitrarily with respect to a temperature gradient using an approximate method. In [101] they treated the motion of a gas bubble near a plane surface due to a temperature gradient oriented arbitrarily with respect to the surface. This was done using the O'Neill solution for non-axisymmetric problems.

The case of two drops oriented arbitrarily with respect to an applied temperature gradient was analyzed by Anderson [102]. He used the method of reflections to calculate the velocity of each drop. He confirmed agreement with the results of Meyyappan *et al.* [98] in the gas bubble limit. Anderson went on to use his solution to predict the thermocapillary migration velocity of a collection of equal sized drops. The motion of such a collection is also the topic of an article by Acrivos *et al.* [103] who point out some interesting similarities to electrophoretic migration problems.

To summarize the findings in the above articles, the interactions of bubbles executing thermocapillary migration either with other bubbles or with neighboring surfaces is weaker than in the case where motion is driven by a body force such as gravity. In low Reynolds number motion, the disturbance fields due to the motion of a bubble differ in the way they decay as one moves away from the bubble depending on the origin of the motion. For gravity driven motion, the disturbance decays as $1/r$ where r is the distance from the bubble. In contrast, for an isolated bubble undergoing thermocapillary motion, the decay occurs as $1/r^3$ which is much more rapid. The temperature gradient disturbance fields also decay as $1/r^3$. Technically, when a surface breaks the fore-aft symmetry in the problem, as mentioned earlier, the decay of the velocity disturbance will occur as $1/r^2$. However, the dominant flow disturbance will decay more rapidly as $1/r^3$ until the bubble is close to the surface. Therefore, when a bubble is approximately three radii away from a plane surface, it moves virtually as though it is isolated. In contrast, for body force driven motion, the effect of a plane surface is felt even when a bubble is ten radii away. Similar results apply when bubbles interact with each other.

Interestingly, Meyyappan *et al.* [98] found that the velocity of the larger of the two bubbles is reduced very slightly whereas that of the smaller bubble is substantially increased over the value it would have if isolated. Thus, they found that two bubbles of equal size move exactly with the velocity each would have if isolated. This fascinating result, uncovered in [98] for the axisymmetric case from numerical evaluation of the analytical result for the bubble velocity, was found to hold good even for arbitrary orientations of the bubbles with respect to the temperature gradient as found by Meyyappan and Subramanian in [100]. For the axisymmetric case, an analytical derivation has now been given by Feuillebois [104]. While the resistance of the surrounding fluid to the motion of each bubble is reduced by the motion of the other (as in similar body force driven motion problems), the **driving force** for the motion, namely the temperature variation over the surface of each bubble, is reduced in such a way as to result in an exact cancellation of the two effects. The above result is shown to be exact in the case of a collection of equal sized bubbles by Acrivos *et al.* [103].

The theory developed by Meyyappan *et al.* [97] for a gas bubble moving normal to a plane surface was extended recently by Barton and Subramanian [88] to the case of a liquid drop. In doing so, they used and extended a technique developed by Sadhal [105] to analytically obtain a set of integrals which were calculated numerically by Meyyappan *et al.* [97]. Their most interesting observation was that when the thermal conductivity ratio of the drop phase to that of the continuous phase is large, the drop can actually move more rapidly near the surface than when it is far away. This counterintuitive result has a logical explanation. Unlike body force driven motion, in thermocapillary migration, the driving force for the motion of the drop can change as noted earlier. For a highly conducting drop, the presence of the plane surface, which is an isotherm in the model, leads to a significant increase of the temperature difference across the drop compared to the case when the drop is isolated. This is sufficient to overcome the increased resistance to motion offered by the neighboring surface.

Merritt and Subramanian [106] have recently published results from experiments on bubbles interacting with a solid plane surface under the combined action of gravity and a temperature gradient. The bubbles were subjected to a downward temperature gradient in the vicinity of both the top and bottom surfaces of the same cell in which isolated bubble experiments had been conducted earlier [46]. As noted before, the bubbles changed in size during the experiments. However, the results were in excellent agreement with theory. The most interesting observation made in [106] was that bubbles near either surface were found to move more rapidly downward than they would if isolated. In this situation, for motion due to gravity, one would predict reduced speed compared to the case of an isolated bubble, and similarly, for motion due purely to thermocapillarity, the bubble should have been retarded by the surface as well. However, the gravitational force was acting to cause upward motion while the thermocapillary force was doing the opposite. The disturbance flows from motion caused by these two types of driving forces decay at different rates as noted earlier. In particular, when a bubble is more than three radii away from the surface, the downward thermocapillary motion should be virtually unaffected since the dominant disturbance flows from this motion decay as $1/r^3$ where r is distance from the bubble center. However, the upward contribution to its velocity from the buoyant force would be considerably retarded, because the disturbance flows from body force driven motion decay as $1/r$. Therefore, the net downward velocity of the bubble is more than the value it would be when the bubble is far away from any surfaces.

Motivated by applications to bubble elimination from space-processed materials, Shankar *et al.* [107] analyzed the motion of a droplet located at the center of a liquid drop when the drop surface is subjected to an arbitrary, but known, axisymmetric temperature distribution. Later, Shankar and Subramanian [108] analyzed the situation wherein a gas bubble is eccentrically located within a liquid drop. In both cases, gravitational effects were neglected, and the drop was assumed to be held fixed by suitable non-contact means such as an acoustic field. In [108] some interesting flow patterns were observed and discussed. Both analyses were restricted to the case of negligible convective transport of momentum and energy.

It can be seen that the literature on interaction problems is limited. The theoretical work is completely confined to the case of negligible convective transport. Also, no theoretical results have been calculated for very small separations for which the solutions in bipolar coordinates are inadequate because of computational difficulties, and lubrication theory would be more useful. Experimental results have only recently been reported for the migration of a gas bubble near a plane surface. K. Barton [109] has performed similar experiments on liquid drops during 1989 as part of his doctoral research, but the results have not yet been published. No experimental or theoretical results are available for conditions when convective transport (and unsteady effects) are important.

This brings us to the conclusion of the discussion of the literature on bubble and drop motion due to interfacial tension gradients. It should be clear that theory is well ahead of experiments in most cases, and there is a great need for more experimental data. Unfortunately, the types of experiments that can be carried out on earth are limited due to strong interference from gravitational effects. Therefore, the next major phase in the accumulation of experimental data will have to await the routine availability of low gravity experimental facilities.

3   Bubble and Drop Migration Induced by Rotation

A gravity-like force can be exerted on suspended objects by inducing rotation of the surrounding fluid. The gyrostatic pressure gradient depends on the density of the fluid involved. Therefore, in the same manner that gravity leads to a net hydrostatic force, rotation will result in a net gyrostatic force which will depend upon the density difference, the angular speed, the size of the object and its distance from the rotation axis. In a given system rotating at a constant angular speed, it is the last factor that differentiates this problem in a fundamental way from that of motion due to gravity wherein the force does not sensibly change with the location of the suspended object within the fluid. When a drop settles due to gravity, we speak of a terminal velocity achieved when the hydrostatic force balances the hydrodynamic drag. In contrast, there is no such terminal velocity in a rotating system due to the continually changing body force on the drop as it moves toward or away from the rotation axis depending upon the sign of the density difference. A drop more dense than the continuous phase will "settle" away from the rotation axis while a less dense drop will move toward the axis. Rigid particles will do likewise, even though we shall focus our attention only on drops and bubbles.

The motion of a rotating fluid appears simple, but is rather intricate. If a container is kept spinning at a constant angular velocity for a sufficiently long period, the entire fluid will ultimately spin at this angular velocity. However, the process by which the fluid reaches this state is complex, and usually involves secondary flows rather than viscous transport of momentum in common fluids. The reader is referred to the treatise by Greenspan [110] and more recent papers on the subject of spin-up for further enlightenment.

Analyses of bubble motion in rotating fluids appeared in [111, 112] stimulated by interest in the movement of bubbles in propellant tanks for rockets. Early experiments were reported by Schrage and Perkins [113] who cited a liquid core nuclear rocket as the application. These investigators injected small gas bubbles into a cylindrical annulus formed by liquid rotating about a horizontal axis. High speed cinematography was used to record the spiral trajectories of the bubbles as they moved toward the core of the annulus. Schrage and Perkins constructed a theoretical description of the migration process using the equations of particle dynamics. Hydrodynamic drag exerted on the bubbles was accommodated by using results good for steady motion of bubbles in large liquid bodies, and the acceleration of the surrounding liquid was included by using the concept of virtual mass. The authors found numerical solutions of their model equations were in agreement with their observations.

Since the appearance of the above articles, Cole and his co-workers have performed a detailed study of bubble and drop motion in fluid contained in a sphere and in a cylinder rotated about a horizontal axis [114, 115, 116, 117]. These authors were motivated by applications to bubble centering inside liquid drops and the management of bubbles and drops in low gravity. The principal changes from the experiments of Schrage and Perkins included the starting of rotation from a state of rest, the use of spherical containers, the study of liquid drops as well as gas bubbles, and in general, the study of a wider range of operating conditions. Some new theoretical descriptions were given, and substantial experimental data were reported.

Cole and coworkers observed that when the liquid is spun up from rest, secondary flows associated with this spin-up process influence the motion of the suspended bubbles and drops. In fact, Annamalai and Cole [115] found that drops slightly more dense than the surrounding liquid can be forced against their natural tendency to move away from the rotation axis by the secondary flows. They found such drops to move initially toward the rotation axis. After a while, the drops moved away from the axis, presumably due to the natural gyrostatic forces after the spin-up flows subsided.

In Europe, Siekmann and coworkers [118, 119, 120, 121] have studied the motion of bubbles in rotating liquids. Both theoretical and experimental work was performed. In the experiments, nearly cylindrical bubbles were confined between two horizontal disks rotated about a vertical axis with the object of minimizing gravitational effects on the migration process. Applications cited by Siekmann and coworkers include the management of bubbles in materials processing in space. A summary is given in [121].

While no simple formulae for bubble speed as a function of the various parameters can be reproduced here, the interested reader is referred to the above articles and the references cited therein for further details.

4   Concluding Remarks

I cannot claim that the material presented in this chapter is comprehensive in its coverage of the state of the literature on bubble and drop motion in reduced gravity. The choice of material reflects my background and bias. However, the reader will recognize that the subject is young and a lot more research needs to be done before a mature state of understanding comparable to that in the case of gravity driven motion of drops and bubbles can be achieved. If this material results in stimulating interest in low gravity fluid mechanics on the part of readers who might be unfamiliar with the subject, it would have achieved one of its principal purposes.

I am indebted to several individuals and institutions for encouragement and support of my own research described herein. I am grateful to Professor William R. Wilcox at Clarkson University for introducing me to the subject of low gravity bubble and drop motion, and for his steady encouragement. Professor Robert Cole, also at Clarkson, has played a key role in stimulating my ventures into the experimental arena, while Professor Michael C. Weinberg, at the University of Arizona, has provided encouragement as well as a steady supply of rewarding scientific discussions over the years. Dialog with my graduate students has been invaluable, and has always served to educate me and broaden my horizons for which I shall forever remain in their debt. In particular, special thanks are due to Kelly D. Barton and David S. Morton both of whom have helped me greatly in the preparation of this chapter. More recently, I have found interactions with Drs. R. Balasubramaniam and Loren H. Dill, both from NASA, very rewarding. Steady financial support for my work on this subject has been provided by NASA's Microgravity Science and Applications Division for which I am most grateful.

5 Nomenclature

Ca Capillary Number defined in Equation (6)

C_p specific heat of the continuous phase

g magnitude of the acceleration due to gravity

k thermal conductivity of the continuous phase

M defined in Equation (12)

Ma Marangoni Number defined in Equation (3)

$P_n(s)$ Legendre Polynomial of degree n and argument s

r distance measured from the center of the drop or bubble

R radius of the drop or bubble

Re Reynolds Number defined in Equation (2)

s $\cos(\theta)$

∇T_∞ temperature gradient in the undisturbed fluid

U scaled velocity of the drop or bubble; $U = \frac{v}{v_o}$

v velocity of the drop or bubble

v_o reference velocity defined in Equation (4)

Wb Weber Number defined in Equation (7)

Z function defined in Equation (10)

α viscosity ratio; $\alpha = \frac{\mu'}{\mu}$

β thermal conductivity ratio; $\beta = \frac{k'}{k}$

γ volumetric heat capacity ratio; $\gamma = \frac{\beta}{\lambda} = \frac{\rho' C_p'}{\rho C_p}$

θ polar angle measured from the direction of drop or bubble motion

κ thermal diffusivity of the continuous phase

λ thermal diffusivity ratio; $\lambda = \frac{\kappa'}{\kappa}$

μ dynamic viscosity of the continuous phase

ν kinematic viscosity of the continuous phase

ρ density of the continuous phase

σ interfacial tension

σ_T rate of change of interfacial tension with temperature

$'$ designates quantities within the drop or bubble

References

[1] Naumann, R.J., and Herring, H.W., *Materials Processing in Space: Early Experiments*, National Aeronautics and Space Administration, Washington, D.C., 1980.

[2] Ostrach, S., Low-Gravity Fluid Flows, in *Annual Reviews of Fluid Mechanics*, vol. 14, eds. M. Van Dyke, J.V. Wehausen, and J.L. Lumley, pp. 313-345, Annual Reviews, California, 1982.

[3] Clift, R., Grace, J.R., and Weber, M.E., *Bubbles, Drops, and Particles*, Academic Press, New York, 1978.

[4] Myshkis, A.D., Babskii, V.G., Kopachevskii, N.D., Slobozhanin, L.A., and Tyuptsov, A.D., *Low-Gravity Fluid Mechanics*, Springer-Verlag, Berlin, 1987.

[5] Wozniak, G., Siekmann, J., and Srulijes, J., Thermocapillary Bubble and Drop Dynamics Under Reduced Gravity - Survey and Prospects, *Z. Flugwiss. Weltraumforsch.*, vol. 12, pp. 137-144, 1988.

[6] Gelles, S.H., and Markworth, A.J., Agglomeration in Immiscible Liquids, SPAR II - Final Report, NASA TM 78125, 1977.

[7] Ahlborn, H., and Löhberg, K., Aluminum-Indium Experiment Soloug - A Sounding Rocket Experiment on Immiscible Alloys, 17th AIAA Conference, New Orleans, Paper No. 79-0172, 1979.

[8] Gelles, S.H., Markworth, A.J., and Mobley, C.E., Low-Gravity Experiments on Liquid Phase Miscibility Gap (LPMG) Alloys - Materials Experiments Assembly (MEA), *Proc. 4th European Symposium on Materials Sciences under Microgravity*, Madrid, Spain, pp. 307-312, 1983.

[9] Walter, H.U., Binary Systems with Miscibility Gap in the Liquid State, in *Materials Sciences in Space*, Eds: B. Feuerbacher, H. Hamacher, and R.J. Naumann, pp. 343-378, Springer-Verlag, Berlin, 1986.

[10] Predel, B., Ratke, L., and Fredriksson, H., Systems With a Miscibility Gap in the Liquid State, in *Fluid Sciences and Materials Science in Space*, Ed. H.U. Walter, pp. 517-565, Springer-Verlag, Berlin, 1987.

[11] Papazian, J.M., and Wilcox, W.R., The Interaction of Bubbles with Solidification Interfaces, *AIAA J.*, vol. 16, pp. 447-451, 1978.

[12] Neilson, G.F., and Weinberg, M.C., Outer Space Formation of a Laser Host Glass, *J. Non-Cryst. Solids*, vol. 23, no. 1, pp. 43-58, 1977.

[13] Uhlmann, D.R., Glass Processing in a Microgravity Environment, in *Materials Processing in the Reduced Gravity Environment of Space*, ed. G.E. Rindone, pp. 269-278, Elsevier, New York, 1982.

[14] Ray, C.S., and Day, D.E., Glass Formation in Microgravity, in *Materials Processing in the Reduced Gravity Environment of Space*, Eds. R.H. Doremus and P.C. Nordine, pp. 239-251, Materials Research Society, Pittsburgh, PA, 1987.

[15] Weinberg, M.C., Glass Processing in Space, *Glass Industry*, vol. 59, pp. 22-27, March 1978.

[16] Annamalai, P., Shankar, N., Cole, R., and Subramanian, R.S., Bubble Migration Inside a Liquid Drop in a Space Laboratory, *Appl. Sci. Res.*, vol. 38, pp. 179-186, 1982.

[17] Subramanian, R.S., Of Drops and Bubbles - The Technology of Space Processing, *Perspectives in Computing*, vol. 4, no. 2/3, pp. 4-19, 1984.

[18] Schmid, L.A., Use of Thermocapillary Migration in a Controllable Heat Valve, *J. Appl. Phys.*, vol. 53, no. 12, pp. 9204-9207, 1982.

[19] Shaw, D.J., *Electrophoresis*, Academic Press, London and New York, 1969.

[20] Saville, D.A., Electrokinetic Effects With Small Particles, in *Annual Reviews of Fluid Mechanics*, vol. 9, eds. M. Van Dyke, J.V. Wehausen, and J.L. Lumley, pp. 321-337, Annual Reviews, California, 1977.

[21] Anderson, J.L., Colloid Transport By Interfacial forces, in *Annual Reviews of Fluid Mechanics*, vol. 21, eds. J.L. Lumley, M. Van Dyke, and H.L. Reed, pp. 61-99, Annual Reviews, California, 1989.

[22] R.L. Rosensweig, Magnetic Fluids, in *Annual Reviews of Fluid Mechanics*, vol. 19, eds. J.L. Lumley, M. Van Dyke, and H.L. Reed, pp. 437-463, Annual Reviews, California, 1987.

[23] Miller, C.A., and Neogi, P., *Interfacial Phenomena: Equilibrium and Dynamic Effects*, pp. 1-23, Marcel Dekker, New York, 1985.

[24] Scriven, L.E., and Sternling, C.V., The Marangoni Effects, *Nature*, vol. 187, pp. 186-188, 1960.

[25] Frumkin, A.N., and Levich, V.G., *Zhur. Fiz. Khim.*, vol. 21, p. 1183, 1947, as cited in Levich, V.G., *Physicochemical Hydrodynamics*, Prentice-Hall, New Jersey, 1962.

[26] Levich, V.G., *Physicochemical Hydrodynamics*, pp. 409-429, Prentice-Hall, New Jersey, 1962.

[27] Savic, P., Circulation and Distortion of Liquid Drops Falling through a Viscous Medium, Natl. Res. Council Canada, Div. Mech. Eng. Rep. MT-22, July 1953.

[28] Davis, R.E., and Acrivos, A., The Influence of Surfactants on the Creeping Motion of Bubbles, *Chem. Eng. Sci.*, vol. 21, pp. 681-685, 1966.

[29] Harper, J.F., The Motion of Bubbles and Drops Through Liquids, *Adv. Appl. Mech.*, vol. 12, pp. 59-129, 1972.

[30] Holbrook, J.A., and LeVan, M.D., Retardation of Droplet Motion by Surfactant. Part 1. Theoretical Development and Asymptotic Solutions, *Chem. Eng. Commun.*, vol. 20, pp. 191-207, 1983.

[31] Holbrook, J.A., and LeVan, M.D., Retardation of Droplet Motion by Surfactant. Part 2. Numerical Solutions For Exterior Diffusion, Surface Diffusion, and Adsorption Kinetics, *Chem. Eng. Commun.*, vol. 20, pp. 273-290, 1983.

[32] LeVan, M.D., and Holbrook, J.A., Motion of a Droplet Containing Surfactant, *J. Colloid Interface Sci.*, vol. 131, no. 1, pp. 242-251, 1989.

[33] Chang, L.S., and Berg, J.C., The Effect of Interfacial Tension Gradients on the Flow Structure of Single drops or Bubbles Translating in an Electric Field, *AIChE J.*, vol. 31, no. 4, pp. 551-557, 1985.

[34] Sadhal, S.S., and Johnson, R.E., Stokes Flow Past Bubbles and Drops Partially Coated With Thin Films. Part 1. Stagnant Cap of Surfactant Film - Exact Solution, *J. Fluid Mech.*, vol. 126, pp. 237-250, 1983.

[35] Quintana, G., Effect of Surfactants on Flow and Mass Transport to Drops and Bubbles, in *Transport Processes in Bubbles, Drops, and Particles*, eds. R.P. Chaabra and D. De Kee, Hemisphere, New York, 1990.

[36] Harper, J.F., Moore, D.W., and Pearson, J.R.A., The Effect of the Variation of Surface Tension with Temperature on the Motion of Bubbles and Drops, *J. Fluid Mech.*, vol. 27, pp. 361-366, 1967.

[37] Levich, V.G., and Krylov, V.S., Surface-Tension-Driven Phenomena, in *Annual Reviews of Fluid Mechanics*, vol. 1, eds. W.R. Sears and M. Van Dyke, pp. 293-316, Annual Reviews, California, 1969.

[38] Schwabe, D., Surface-Tension-Driven Flow in Crystal Growth Melts, in *Crystals*, pp. 75-112, Springer-Verlag Berling, Heidelberg, 1988.

[39] Young, N.O., Goldstein, J.S., and Block, M.J., The Motion of Bubbles in a Vertical Temperature Gradient, *J. Fluid Mech.*, vol. 6, pp. 350-356, 1959.

[40] Chifu, E., Stan, I., Finta, Z., and Gavrila, E., Marangoni-Type Surface Flow on an Undeformable Free Drop, *J. Colloid Interface Sci.*, vol. 93, no. 1, pp. 140-150, 1983.

[41] Ruckenstein, E., Can Phoretic Motions be Treated as Interfacial Tension Gradient Driven Phenomena?, *J. Colloid Interface Sci.*, vol. 83, no. 1, pp. 77-81, 1981.

[42] Trefethen, L.M., Surface Tension in Fluid Mechanics, a color film by Encyclopedia Brittanica Educational Corporation, Film No. 21610, 1963.

[43] Kuznetsov, V.M., Lugovtsov, B.A., Sher, Y.I., Motion of Gas Bubbles in a Liquid Under the Influence of a Temperature Gradient, *Zh. Prikl. Mekh. Tekh. Fiz.*, pp. 124-126, Jan.-Feb. 1966; Original in Russian, NASA Technical Translation NASA TT F-16569, September 1975.

[44] Hardy, S.C., The Motion of Bubbles in a Vertical Temperature Gradient, *J. Colloid Interface Sci.*, vol. 69, no. 1, pp. 157-162, 1979.

[45] Merritt, R.M., Bubble Migration and Interactions in a Vertical Temperature Gradient, Ph. D. Thesis, Clarkson University, 1988.

[46] Merritt, R.M., and Subramanian, R.S., The Migration of Isolated Gas Bubbles in a Vertical Temperature Gradient, *J. Colloid Interface Sci.*, vol. 125, no. 1, pp. 333-339, 1988.

[47] McGrew, J.L., Rehm, T.L., and Griskey, R.G., The Effect of Temperature Induced Surface Tension Gradients on Bubble Mechanics, *Appl. Sci. Res.*, vol. 29, pp. 195-210, 1974.

[48] Wilcox, W.R., Subramanian, R.S., Papazian, J.M., Smith, H.D., and Mattox, D.M., Screening of Liquids for Thermocapillary Bubble Movement, *AIAA J.*, vol. 17, no. 9, pp. 1022-1024, 1979.

[49] Mattox, D.M., Smith, H.D., Wilcox, W.R., and Subramanian, R.S., Thermal-Gradient-Induced Migration of Bubbles in Molten Glass, *J. Amer. Ceram. Soc.*, vol. 65, no. 9, pp. 437-442, 1982.

[50] Smith, H.D., Mattox, D.M., Wilcox, W.R., Subramanian, R.S., and Meyyappan, M., Experimental Observation of the Thermocapillary Driven Motion of Bubbles in a Molten Glass Under Low Gravity Conditions, in *Materials Processing in the Reduced Gravity Environment of Space*, ed. G.E. Rindone, pp. 279-288, Elsevier, New York, 1982.

[51] Meyyappan, M., Subramanian, R.S., Wilcox, W.R., and Smith, H.D., Bubble Behavior in Molten Glass in a Temperature Gradient, in *Materials Processing in the Reduced Gravity Environment of Space*, ed. G.E. Rindone, pp. 311-314, Elsevier, New York, 1982.

[52] Papazian, J.M., Gutowski, R., and Wilcox, W.R., Bubble Behavior During Solidification in Low Gravity, *AIAA J.*, vol. 17, no. 10, pp. 1111-1117, 1979.

[53] Chandrasekhar, S., *Hydrodynamic and Hydromagnetic Stability*, pp. 9-75, Oxford University Press, Oxford, 1961.

[54] Thompson, R.L., Marangoni Bubble Motion in Zero Gravity, Ph.D. Thesis, University of Toledo, 1979.

[55] Thompson, R.L., Dewitt, K.J., and Labus, T.L., Marangoni Bubble Motion Phenomenon in Zero Gravity, *Chem. Eng. Commun.*, vol. 5, pp. 299-314, 1980.

[56] Crespo, A., and Manuel, F., Bubble Motion Under Reduced Gravity, *Proc. 4th European Symposium on Materials Sciences under Microgravity*, Madrid, Spain, pp. 45-49, 1983.

[57] Balasubramaniam, R., and Chai, A.T., Thermocapillary Migration of Droplets: An Exact Solution for Small Marangoni Numbers, *J. Colloid Interface Sci.*, vol. 119, no. 2, pp. 531-538, 1987.

[58] Nähle, R., Neuhaus, D., Siekmann, J., Wozniak, G., and Srulijes, J., Separation of Fluid Phases and Bubble Dynamics in a Temperature Gradient - A Spacelab D1 Experiment, *Z. Flugwiss. Weltraumforsch.*, vol. 11, pp. 211-213, 1987.

[59] Szymczyk, J.A., Wozniak, G., and Siekmann, J., On Marangoni Bubble Motion at Higher Reynolds and Marangoni Numbers Under Microgravity, *Appl. Microgravity Tech.*, vol. I, no. 1, pp. 27-29, 1987.

[89] Morton, D.S., and Subramanian, R.S., The Migration of a Compound Drop Due to Thermocapillarity II. Eccentric Case, Submitted for Publication, Physics of Fluids A, 1989.

[90] Stimson, M., and Jeffery, G.B., The Motion Of Two Spheres in a Viscous Fluid, *Proc. R. Soc.*, vol. A111, pp. 110-116, 1926.

[91] Rushton, E., and Davies, G.A., The Slow Unsteady Settling of Two Fluid Spheres Along Their Line of Centers, *Appl. Sci. Res.*, vol. 28, pp. 37-61, 1973.

[92] Rushton, E., and Davies, G.A., The Slow Motion of Two Spherical Particles Along Their Line of Centers, *Int. J. Multiphase Flow*, vol. 4, 357-381, 1978.

[93] Haber, S., Hetsroni, G., and Solan, A., On the Low Reynolds Number Motion of Two Droplets, *Int. J. Multiphase Flow*, vol. 1, pp. 57-71, 1973.

[94] Brenner, H., The Slow Motion of a Sphere Through a Viscous Fluid Towards a Plane Surface, *Chem. Eng. Sci.*, vol. 16, pp. 242-251, 1961.

[95] Bart, E., The Slow Unsteady Settling of a Fluid Sphere Toward a Flat Fluid Interface, *Chem. Eng. Sci.*, vol. 23, pp. 193-210, 1968.

[96] O'Neill, M.E., A Slow Motion of Viscous Liquid Caused by a Slowly Moving Solid Sphere, *Mathematika*, vol. 11, pp. 67-74, 1964.

[97] Meyyappan, M., Wilcox, W.R., and Subramanian, R.S., Thermocapillary Migration of a Bubble Normal to a Plane Surface, *J. Colloid Interface Sci.*, vol. 83, no. 1, pp. 199-208, 1981.

[98] Meyyappan, M., Wilcox, W.R., and Subramanian, R.S., The Slow Axisymmetric Motion of Two Bubbles in a Thermal Gradient, *J. Colloid Interface Sci.*, vol. 94, no. 1, pp. 243-257, 1983.

[99] Jeffery, G.B., On a Form of the Solution of Laplace's Equation Suitable for Problems Relating to Two Spheres, *Proc. R. Soc.*, vol. A87, pp. 109-120, 1912.

[100] Meyyappan, M., and Subramanian, R.S., The Thermocapillary Motion of Two Bubbles Oriented Arbitrarily Relative to a Thermal Gradient, *J. Colloid Interface Sci.*, vol. 97, no. 1, pp. 291-294, 1984.

[101] Meyyappan, M., and Subramanian, R.S., Thermocapillary Migration of a Gas Bubble in an Arbitrary Direction with Respect to a Plane Surface, *J. Colloid Interface Sci.*, vol. 115, no. 1, pp. 206-219, 1987.

[102] Anderson, J.L., Droplet Interactions in Thermocapillary Motion, *Int. J. Multiphase Flow*, vol. 11, no. 6, pp. 813-824, 1985.

[103] Acrivos, A., Jeffrey, D.J., and Saville, D.A., Thermocapillary and Electrophoretic Migration of Particles in Suspension, *J. Fluid Mech.*, in press, 1990.

[104] Feuillebois, F., Thermocapillary Migration of Two Equal Bubbles Parallel to Their Line of Centers, *J. Colloid Interface Sci.*, vol. 131, no. 1, pp. 267-274, 1989.

[105] Sadhal, S.S., A Note on the Thermocapillary Migration of a Bubble Normal to a Plane Surface, *J. Colloid Interface Sci.*, vol. 95, no. 1, pp. 283-285, 1983.

[106] Merritt, R.M., and Subramanian, R.S., Migration of a Gas Bubble Normal to a Plane Horizontal Surface in a Vertical Temperature Gradient, *J. Colloid Interface Sci.*, vol. 131, no. 2, pp. 514-525, 1989.

[107] Shankar, N., Cole, R., and Subramanian, R.S., Thermocapillary Migration of a Fluid Droplet Inside a Drop in a Space Laboratory, *Int. J. Multiphase Flow*, vol. 7, no. 6, pp. 581-594, 1981.

[108] Shankar, N., and Subramanian, R.S., The Slow Axisymmetric Thermocapillary Migration of an Eccentrically Placed Bubble Inside a Drop in Zero Gravity, *J. Colloid Interface Sci.*, vol. 94, no. 1, pp. 258-275, 1983.

[109] Barton, K.D., Ph.D. Thesis in Chemical Engineering, Clarkson University, 1990.

[110] Greenspan, H.P., *The Theory of Rotating Fluids*, Cambridge University Press, London, England, 1969.

[111] Clifton, J.V., Hopkins, R.A., and Goodwin, D.W., Motion of a Bubble in a Rotating Liquid, *J. Spacecraft*, vol. 6, no. 2, pp. 215-217, 1969.

[112] Catton, I., and Schwartz, S.H., Motion of Bubbles in a Rotating Container, *J. Spacecraft*, vol. 9, no. 6, pp. 468-471, 1972.

[113] Schrage, D.L., and Perkins, Jr., H.C., Isothermal Bubble Motion through a Rotating Liquid, *Trans. ASME J. Basic Eng.*, vol. 94, pp. 187-192, 1972.

[114] Annamalai, P., Subramanian, R.S., and Cole, R., Bubble Migration in a Rotating, Liquid-Filled Sphere, *Phys. Fluids*, vol. 25, no. 7, pp. 1121-1126, 1982.

[115] Annamalai, P., and Cole, R., Drop Motion in a Rotating Immiscible Liquid Body, in *Advances in Space Research*, ed. Y. Malmejac, vol. 3, no. 5, pp. 165-168, Pergamon Press, Great Britain, 1983.

[116] Annamalai, P., and Cole, R., Particle Migration in Rotating Liquids, *Phys. Fluids*, vol. 29, no. 3, pp. 647-649, 1986.

[117] Ruggles, J.S., Cook, R.G., Annamalai, P., and Cole, R., Bubble and Drop Trajectories in Rotating Flows, *Experimental Thermal and Fluid Science*, vol. 1, pp. 293-301, 1988.

[118] Siekmann, J., and Johann, W., On Bubble Motion in a Rotating Liquid Under Simulated Low and Zero Gravity, *Ingenieur-Archiv*, vol. 45, pp. 307-315, 1976.

[119] Siekmann, J., and Dittrich, K., Note on Bubble Motion in a Rotating Liquid Under Residual Gravity, *Int. J. Non-Linear Mechanics*, vol. 12, pp. 409-415, 1977.

[120] Siekmann, J., and Dittrich, K., Computer Study of Bubble Motion in a Rotating Liquid, *Comp. Meth. App. Mech. Eng.*, vol. 10, pp. 291-301, 1977.

[121] Siekmann, J., Migration of Gas Bubbles Under Various Force Fields in a Microgravitational Environment, *Proc. 3rd European Symposium on Materials Sciences in Space*, Grenoble, France, pp. 341-347, 1979.

Chapter 2

Fluid Particles in Rheologically Complex Media

R. P. CHHABRA
Department of Chemical Engineering
Indian Institute of Technology
Kanpur 208016, India

D. DE KEE
Department of Chemical Engineering
University of Sherbrooke
Sherbrooke, Quebec, J1K 2R1, Canada

INTRODUCTION

The formation and motion of bubbles in liquids is of considerable industrial importance in a range of chemical and processing industries. Typical examples of where interactions between bubbles and a liquid phase are encountered include bubble columns, involving gas holdup and backmixing and stirred vessels for effecting gas-liquid reactions, wastewater treatment, the production of foams and batters, the behaviour of gas bubbles in blood, subjected to oscillating pressure fields and possibly to acoustic effects, etc. Furthermore, the interphase transfer between a gas bubble and a continuous phase plays the central role in determining the performance of chemical reactors. Another associated phenomenon encountered inevitably in multibubble systems is that of coalescence and breakage of bubbles. Coalescence in air-sparged bioreactors is of particular interest. Depending upon the envisaged application, coalescence may be desirable (e.g. in promoting separation) or detrimental to a process (e.g. gas-liquid reactions where a large interfacial area is required). One would intuitively expect the process of coalescence to be influenced by a large number of process and physical variables including shapes and size of bubbles, physical properties of the two phases (density, viscosity, surface tension, etc.), the presence of surface active agents and the prevailing flow patterns, e.g. uniform, shear or elongational flow fields, etc. Furthermore, often swarms of gas bubbles are encountered in real life applications. Owing to the inherent complex nature of bubble phenomena coupled with the large number of influencing variables, a completely theoretical analysis is virtually impossible. On the other hand, it is readily acknowledged that the behaviour of single bubbles can provide useful insights into the mysterious world of the multi-bubble situations. In view of this, this chapter is primarily concerned with the fluid mechanics of single bubbles in rheologically complex media. Extensions to multiparticle situations are suggested wherever possible.

Over the years, a wealth of information has accumulated concerning the formation, motion and coalescence of bubbles in Newtonian fluids. Consequently, a reasonably coherent picture has emerged. Excellent accounts of the developments in this area are available in the literature [Clift et al., 1978; Hetsroni, 1982; Kumar and Kuloor, 1970; Marrucci, 1969; Narayanan et al., 1974; Blass, 1990, etc.].

Unfortunately, not all liquids of industrial importance display simple Newtonian characteristics. Indeed, in some of the aforementioned applications, the liquid phase exhibits a range of rheological complexities including shear rate dependent viscosity, viscoelasticity, time dependent behaviour, etc. Typical examples involving bubbles in rheologically complex media include fermentation broths and other biotechnological systems, manufacture of paints and detergents, drying of food products, etc. and others as have been pointed out by Mashelkar [1980], Blanch and Bhavaraju [1976[, Kemblowski and Kristiansen [1986] and Kelkar and Shah [1985].

Despite the overwhelming pragmatic importance of this flow configuration, it has not only received a very scant attention but more importantly even the resulting picture is very confused and incoherent. In addition to its practical relevance, even from a purely theoretical viewpoint, the motion of a single bubble constitutes an interesting example of a strongly non-viscometric flow wherein major macroscopic differences have been observed from its Newtonian counterpart.

This chapter aims to provide a comprehensive review of the developments in this area in general; however, special attention would be given to the following aspects of bubble phenomena in rheologically complex media:

(1) formation of bubbles in stagnant and flowing liquids,
(2) shapes of bubbles,
(3) terminal rise velocity of single and multiple bubbles in stationary liquids,
(4) coalescence under idealized conditions.

FORMATION OF SINGLE BUBBLES FROM SUBMERGED ORIFICES IN STAGNANT LIQUIDS

There are numerous ways of effecting gas-liquid contacting; perhaps the simplest, and possibly also the most widely used method is the dispersion of a gas through submerged nozzles, slots, and holes. Thus, most of the efforts have been directed at investigating the formation of single bubbles from submerged orifices. There is no question that the parameter of central interest is the size of the produced bubble (volume/diameter) under specified conditions, for a given gas-liquid system.

A general model encompassing the formation of bubbles under all conditions is not yet available, even for Newtonian liquids. All models have thus inherent varying levels of approximations. There are essentially two models which have gained wide acceptance in the literature [Davidson and Schuler, 1960; Kumar and Kuloor, 1970]. In contrast, not only is there very little known about the formation of bubbles in non-Newtonian media but most of the developments in this area can be regarded as straight forward extensions of the aforementioned models. It is therefore regarded to be instructive and desirable to recapitulate the salient features of both these models.

DAVIDSON-SCHULER MODEL

In this model, the bubble is assumed to form at a point source where the gas is supplied. As the bubble is formed, it moves upward with a velocity determined by the resultant of various forces acting on it. The detachment is assumed to occur when the centre of the bubble has moved by the distance equal to the sum of the radius of the orifice and the bubble. Figure 1 shows this idealized process. Furthermore, during the whole formation period, the bubble is assumed to be spherical. Depending upon the physical properties of the liquid phase (density, viscosity, surface tension) and the flow rate of the gas, Davidson and Schuler [1960] have identified various regimes of bubble formation. For instance, in low viscosity systems, the flow can be assumed to be irrotational and furthermore if the surface tension effects are negligible, the only relevant forces acting on the bubble are that due to buoyancy and due to the inertia of the liquid moving with the expanding bubble, viz:

$$V\left(\rho_L - \rho_g\right)g = \frac{d}{dt}\left\{\left(\frac{11}{16}\rho_L + \rho_g\right)V \cdot \frac{dx}{dt}\right\}$$

(1)

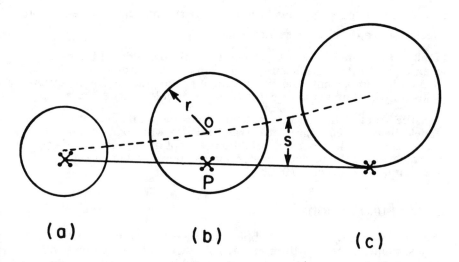

(a)   (b)   (c)

FIGURE 1.   Idealized sequence of bubble formation in a liquid - Davidson and Schuler model (1960).

On the right side of Equ. 1, the factor (11/16) ρ_L accounts for the virtual mass of the liquid moving along with the bubble. Under constant gas flowrate (Q), and satisfying the initial conditions x=0 and

$\frac{dx}{dt} = 0$ at t=0, one can easily show that the volume of a bubble at detachment is given by:

$$V_f = 1.378 \, Q^{6/5} \, g^{-3/5} \tag{2}$$

Subsequently, Davidson and Harrison [1963] have replaced the factor of (11/16) by (1/2) in Equ. (1). Likewise, in the case of highly viscous systems, at appropriate flowrates, the inertial force of the liquid being carried by the gas bubble would be negligible and instead the buoyancy force will be counterbalanced by the viscous drag force. Davidson and Schuler [1960] argued that at sufficiently low flowrates, the drag force can be approximated by the Stokes drag force. This leads to the following simple expression for the bubble volume at the time of detachment:

$$V_f = \left(\frac{4\pi}{3}\right)^{1/4} \left(\frac{15\mu Q}{2\rho_L g}\right)^{3/4} \tag{3}$$

Attention is drawn to the fact that in arriving at Equs. (2) and (3) the approximation $\rho_g << \rho_L$ has been tacitly used, and the surface tension effects have altogether been neglected.

Finally, Davidson and Schuler [1960] have also considered the case when the flowrate is sufficiently high so as to make the inertial force contribution important, and the inclusion of the latter in the analysis yields a rather cumbersome expression for bubble volume.

Over the years, it has been established that the predictions of this model show a good degree of correspondence with experimental results under wide ranges of conditions except at extremely low gas flowrates where the surface tension forces appear to play a significant role [Kumar and Kuloor, 1970].

KUMAR-KULOOR MODEL

This approach envisions the bubble formation to take place in two stages, namely, the growth or expansion stage and the detachment stage. Figure 2 shows schematically the bubble formation according to this model. It differs from the model of Davidson and Schuler in so far that the bubble stays at the orifice tip during the growth stage

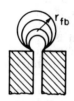  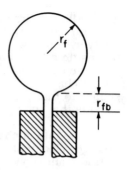

Expansion stage   Detachment stage   Condition of detachment

FIGURE 2. Idealized sequence of bubble formation - Kumar and Kuloor
model (1970).

whereas in the second stage it moves away from (but remains in
contact with) the orifice tip via a neck until detachment. During the
second stage, the bubble not only moves away from the orifice but also
keeps expanding due to the continuous gas flow. It should be recognized that a bubble will only lift off from the orifice tip when
there is a net upward force acting on it. Kumar and Kuloor [1970]
have theorized these two stages as follows; the forces acting on a
bubble during the first stage are: buoyancy, in the upward direction
and surface tension and drag force acting downward. Thus, the first
stage is assumed to end when the net force acting on the bubble is
zero. The second stage is characterized mainly by the motion of an
expanding bubble which can be described by Newton's law of motion.
Finally, detachment is assumed to occur when the bubble has travelled
a distance equal to its radius at the end of the first stage. In other
words, the final bubble volume contains two components as follows:

$$V_f = V_{fb} + Qt_c$$

(4)

where t_c is the time of detachment, and V_{fb} is the bubble volume at
the end of the first stage.

In this model therefore the problem reduces to that of finding the
bubble volume at the end of the first stage. This is evaluated again by
writing a force balance. In the absence of surface tension effects,
there are essentially three forces acting on the bubble: buoyancy

(upward), and drag and inertial forces (acting vertically downward). One can again invoke the irrotational flow approximation for low viscosity systems, whereas the situation can be approximated by the Stokes drag force for viscous systems. The inertial force resulting from the expansion of the bubble can be expressed in terms of the rate of change of momentum of the forming bubble, in a manner similar to that presented by Davidson and Schuler [1960]. It is obvious that the upper part of the bubble moves with a velocity equal to its rate of change of diameter, whereas the base remains stationary. Thus, the rate of change of momentum is evaluated by using the average velocity at the centre of the bubble. The force balance written in this manner yields the following implicit expression for the bubble volume at the end of the first stage:

$$V_{fb} = 0.0474 \frac{Q^2 V_{fb}^{-2/3}}{g} + 2.42 \left(\frac{\mu}{\rho_L}\right) V_{fb}^{-1/3} \left(\frac{Q}{g}\right) \tag{5}$$

The equation of motion for the bubble centre can now be written as:

$$\frac{d}{dt}(MV') = \left(V_{fb} + Qt_c\right)\left(\rho_L - \rho_g\right)g - 6\pi r \mu V' \tag{6}$$

where V' is the velocity of the bubble centre (i.e. dr/dt) plus the velocity of the base. Substitution for various terms in Equation (6), e.g.

$M = Qt_c \left((11/16) \rho_L + \rho_g\right)$, etc., and using the condition of detachment as the distance travelled equal to the radius of the bubble at the end of the first stage yields the required expression for bubble volume at the time of detachment. Kumar and Kuloor [1970] have deduced several special cases, and have also used the same reasoning to describe drop formation.

It is somewhat surprising that despite their widely different physical background, there is a striking similarity in the expression for bubble volume under inviscid flow conditions, i.e., both models reduce to:

$$V_f = C \left(\frac{Q^2}{g}\right)^{3/5} \tag{7}$$

In Equation (7), C is a constant. Davidson and Schuler [1960] found its value to be 1.387. This was subsequently modified to 1.138 [Davidson and Harrison, 1963]. The two stage model of Kumar and Kuloor yields a value of 0.976. All these values compare favourably with the experimental value of 1.722 [Krevelen and Hoftiyzer, 1950]. Detailed comparisons between the predictions of the two models with experiments show that the two stage model of Kumar and Kuloor [1970] performs better than that of Davidson and Schuler [1960], particularly under those conditions where the surface tension effects are appreciable. An excellent review on bubble and drop formation is available in the literature [Kumar and Kuloor, 1970].

BUBBLE FORMATION IN NON-NEWTONIAN MEDIA

One would intuitively expect that either of the two aforementioned models can be adapted for bubble formation in purely viscous liquids whereas additional complications are likely to arise in the case of viscoelastic liquids owing to unequal normal stresses. In the case of purely viscous liquids (e.g. power law fluids, etc.), one can easily modify the drag force on the bubble. The resulting expression, as obtained by Subramaniyan and Kumar [1967], yields:

$$V_{fb}(\rho_L - \rho_g)\, g = \frac{Q^2 \left[\rho_g + \frac{11}{16}\rho_L\right] V_{fb}^{-2/3}}{12\pi(3/4\,\pi)^{2/3}} + \frac{24\, Xk\, Q^n V_{fb}^{\frac{2-3n}{3}}}{2^{3n+1}\pi^{n-1}(3/4\,\pi)^{\frac{3n-2}{3}}}$$

(8)

where n and k are the power law constants and X is a correction factor to be applied to the Stokes formula to obtain the drag on a bubble in a power law liquid. The approximate functional dependence of X on the power law index is available in the literature [Hirose and Moo-Young, 1969; Chhabra and Dhingra, 1986].

In contrast to the extensive results available on bubble formation in Newtonian fluids, there is a real dearth of experimental results on the formation of bubbles in well characterized non-Newtonian media. Most polymer solutions generally used as model non-Newtonian experimental fluids are highly viscous. The inviscid models are therefore likely to be applicable only at extremely high gas flow rates. Indeed, Acharya et al. [1978] have provided a limited amount of data which substantiates this assertion. Their experimental values of bubble volume lie somewhere in between the predictions of the two models (presented earlier) for inviscid conditions. In fact, in this flow regime, even the fluid elasticity was found to exert virtually no influence on the formation or on the volume of the bubbles. At low flowrates, it has generally been argued [Rabiger and Vogelpohl, 1986; Ghosh and Ulbrecht, 1989] that the fluid elasticity significantly influences the shape of the bubbles at the detachment stage. That is: the bubbles are elongated and connected with the orifice tip via a "neck" thereby resulting in detachment times much larger than those anticipated by the models of Davidson and Schuler [1960] and of Kumar and Kuloor [1970]. Based on qualitative photographic evidence, both Rabiger and Vogelpohl [1986] and Ghosh and Ulbrecht [1989] have identified the so-called "waiting" stage in the formation of bubbles. It is however not at all clear, how this behaviour can be incorporated into the existing or future models of bubble formation.

Costes and Alran [1978] reported experimental values of bubble size (i.e. volume) for one power law liquid and concluded that the inviscid theory of Davidson and Schuler [1960] adequately described the limited experimental results. Based on the value of the orifice Reynolds number, they proposed a criterion for delineating the regimes of bubble formation. Miyahara et al. [1988] have also studied

the formation of bubbles in highly viscous non-Newtonian environ-
ments with special reference to the effect of chamber volume. Based
on dimensional considerations, these investigators have outlined a
scheme for delineating the regimes of bubble formation, viz., constant
flow or constant pressure conditions with a transition zone in between
these two limiting behaviours. More recently, Tsuge and Terasaka
[1989] have reported experimental results on the volumes of bubbles
formed at orifices submerged in highly viscous Newtonian and non-
Newtonian liquids. However, no comparisons with the other contem-
porary works are reported. Instead a dimensionless correlation has
been presented.

The formation of bubbles in flowing liquids has been explored by
Kezios and Schowalter [1986] and by Ghosh and Ulbrecht [1989], to
name a few. The former generated bubbles by focussing the energy
discharged from a Q-switched laser into a small volume of dilute
polymer solution undergoing shear between coaxial cylinders. Bubble
formation was associated with a nanosecond time scale, while bubble
growth and collapse occured in a few milliseconds. Non-sphericity
and alteration in cavitation behaviour are discussed. Ghosh and
Ulbrecht related the effect of fluid motion to the alteration of the drag
force, experienced by a growing bubble.

From the foregoing description, it is safe to conclude that not only
very little work has been done on bubble formation in non-Newtonian
media but more importantly the work reported to date has been of an
exploratory nature. This remains an area for future activity.

BUBBLE SHAPES

The shapes of bubbles are primarily governed by the relative magnitudes of inertial, viscous, surface tension, gravity and buoyancy forces.
Elastic forces come into play when dealing with viscoelastic fluids.

In the case of bubbles moving in a constant viscosity (3.3 mPa.s)
aqueous glycerine solution, De Kee and Chhabra [1988] presented
photographs of the observed bubble shapes, under conditions where
the Reynolds number varied from about 300 to 1400. Bubbles as small
as 5.10^{-8} m^3 were shown to be distorted from a spherical shape. It
was further observed, that the bubble shape is not uniquely dependent
upon the bubble volume. In the case of bubbles moving in such
Newtonian liquids, the shape is largely governed by inertial, viscous
and surface tension forces. Based on this, several so called "shape
maps" have been reported in the literature [Clift et al., 1978].
Intuitively, it appears that similar (but adapted for non-Newtonian
fluids) dimensionless groups (representing the ratios of the various
forces mentioned earlier should be useful in describing and predicting
the shapes of the bubbles. Unfortunately, qualitatively different shapes
which are unique to viscoelastic liquids are observed and hence a
reconciliation with the "shape maps" referred to above does not seem
possible. Indeed, at low Reynolds number (i.e., negligible inertial
forces), the fluid viscoelasticity is known to play a major role in
determining the shape of gas bubbles.
Bubbles moving in non-Newtonian and in viscoelastic media retain
their spherical shape up to larger volumes with increasing polymer

50

concentration, i.e. pseudoplasticity and viscoelasticity. Figure 3 depicts schematically the shapes observed in a study dealing with polyacrylamide and with carboxymethyl cellulose solutions [De Kee et al., 1990]. Air bubbles can undergo shape changes from spherical to prolate-teardrop, to oblate-cusped, to oblate and finally to Taylor-Davies type spherical caps, depending on the bubble volume and on the solution properties. Table I relates bubble shape to bubble volume, in the case of a 0.5% solution of polyacrylamide (PAA) in a 50 wt. % mixture of water and glycerine.

FIGURE 3. Observed bubble shapes in pseudoplastic and in viscoelastic solutions [from A. Dajan, 1985].

TABLE I. Bubble shapes in a 0.5% polyacrylamide solution.

Shape	Bubble volume x 10^8 (m^3)
Spherical	< 4
Prolate teardrop	4 - 15
Oblate cusped	15 - 250
Spherical cap	> 250

Fluid elasticity is responsible for shape transitions and other associated phenomena such as tailing, etc.; the latter has been reported by Barnett et al. [1966] who also postulated that the fluid displaced by the ascending bubble did not recover quickly enough, thereby creating a "hole" at the rear of the gas bubble. This, in turn, sets up a negative pressure gradient and a wake is formed. The time lag in recovery is related to the fluid relaxation time, which increases with increasing polymer concentration. Thus, one would intuitively expect the wake behind a bubble to be "larger" in a more viscoelastic liquid than in a mildly viscoelastic one.

Figure 4 illustrates shapes observed in a 1% polyacrylamide solution, in a 50 wt. % mixture of water and glycerine.

As pointed out earlier [De Kee and Chhabra, 1988], shapes do not appear to be influenced to a discernable extent by increasing levels of viscoelasticity. The transition from one shape to another appears somehow to be a function of fluid rheology. That is to say: as the fluid viscoelasticity increases, larger and larger bubbles stay spherical and the transition to the spherical cap region occurs at larger values of bubble volume.

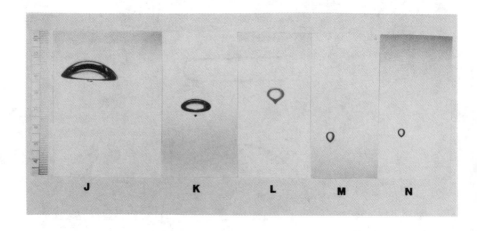

FIGURE 4.   Shapes in a 1% PAA solution (0.12 < Re < 6)
J = 1.10^{-5} m^3, K = 2.10^{-6} m^3, L = 1.10^{-6} m^3
M = 1.10^{-7} m^3, N = 5.10^{-8} m^3.

MOTION OF SINGLE BUBBLES IN NON-NEWTONIAN MEDIA

A considerable body of literature exists on the motion of single and multiple bubbles moving in non-Newtonian media. A brief summary is provided in Table II. There is no question that major research efforts have been directed at elucidating the following two aspects of bubble phenomena in rheologically complex media:

(i)   Terminal rise velocity - volume behaviour, and
(ii) Drag coefficient - Reynolds number relationship.

Detailed consideration will be given to each of these aspects in the ensuing sections.

TERMINAL RISE VELOCITY - VOLUME BEHAVIOUR

Perhaps the most striking phenomenon associated with the free rise of single bubbles in non-Newtonian fluids is the oft reported "discontinuity" or "abrupt transition" in terminal velocity when measured as a function of bubble size. Astarita and Appuzzo [1965] were the first to report this anomalous behaviour and described a 6 to 10 fold increase in terminal velocity at a critical bubble size. Subsequently, similar results have been reported by many other workers [Calderbank et al., 1970; Leal et al., 1971; Acharya et al., 1977, etc.]. Figure 5, replotted from the data of Astarita and Appuzzo [1965], clearly shows the phenomenon. However, it is also appropriate to add here that no such discontinuity has been observed by Macedo and Yang [1974], De Kee et al. [1986; 1990], and by De Kee and Chhabra [1988] (Figure 6, replotted from Dajan [1985], shows no abrupt discontinuity. It is therefore not at all clear whether the phenomenon is a real effect or simply an artifact of the extremely sensitive nature of the experimental work to the level of cleanliness, the contamination, etc.

From our understanding of the motion of bubbles moving in Newtonian fluids, it is readily acknowledged that small bubbles behave like rigid spheres (and follow Stokes's law) whereas large bubbles exhibit a mobile surface which is usually assumed to be shear free. Thus one would expect a change of slope (from 1 to 1.5) even in Newtonian liquids. However, it is not at all clear whether this changeover from "no slip" to "zero shear" boundary condition at the bubble surface would manifest itself as an abrupt transition such as that reported by Astarita and Appuzzo [1965] or in the form of a gradual transition as reported by De Kee et al. [1986, 1990]. Based on a range of considerations, several attempts have been made to predict the critical size for this changeover. Most of these have been reviewed by Chhabra [1988] and by Kawase and Ulbrecht [1981]; the latter authors have also extended the analysis of Bond and Newton [1928] to predict the critical bubble size as follows:

$$R_c = \left(\frac{\sigma}{\Delta\rho g}\right)^{0.5}$$

(9)

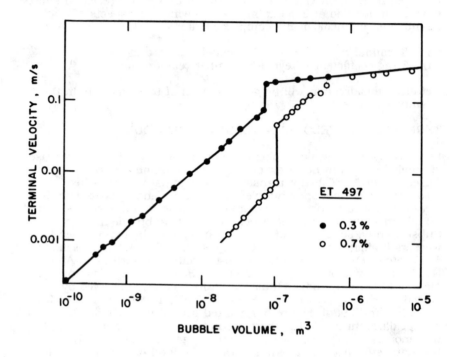

FIGURE 5. Bubble velocity-volume data for air bubbles rising through aqueous solutions of ET497 (replotted from Astarita and Appuzzo).

Preliminary comparisons between the predictions of Eq.9 and the experimental values, encompassing wide ranges of rheological conditions, are encouraging. It is, however, surprising that the liquid rheology plays no role in this process! Zana and Leal [1978] examined the dynamics of a single bubble in viscoinelastic and viscoelastic media, and concluded that the shear rate dependent viscosity can account for about 30% of the increase in velocity at the transition point. The introduction of even a small degree of fluid viscoelasticity (essentially tantamounting to unequal primary normal stresses) is sufficient to explain the observed change in velocity. Undoubtedly, the elegant study of Zana and Leal [1978] has led to a better understanding of the phenomenon but it is not yet refined to the extent to be used for predictive calculations. Thus there is no consensus among various workers on this point.

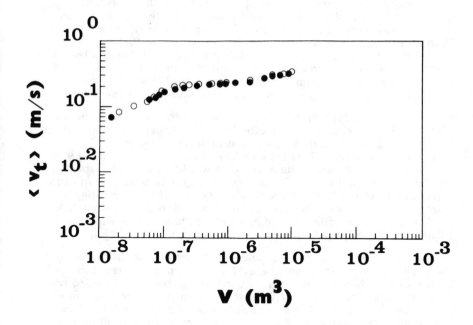

FIGURE 6. Motion of air (O) and nitrogen (●) bubbles in a 0.5% poly-acrylamide solution in a 50 wt. % mixture of water and glycerine (replotted from A. Dajan).

DRAG COEFFICIENT-REYNOLDS NUMBER RELATIONSHIP

In processing operations involving relative motion between gas bubbles and a continuous liquid phase, the residence time is an important variable whose value depends upon the size of the equipment and on the rise velocity of the bubbles. Thus the need to calculate the rise velocity of a single bubble or of an ensemble of bubbles often arises. This information is customarily expressed in dimensionless form with the help of drag coefficient, Reynolds number and other parameters. In the case of Newtonian fluids, the functional dependence of the drag coefficient on the Reynolds and other pertinent dimensionless parameters is well established [Clift et al., 1978].

From a theoretical standpoint, the governing equations describing the motion of a single bubble in non-Newtonian media are highly nonlinear even when the inertial terms are altogether neglected. To date therefore, only approximate solutions have been obtained which are primarily limited to creeping flow situations only. Most of the analyses can be divided into three groups depending upon the mathematical technique employed to seek approximations:

(i) Variational principles have been used extensively in obtaining upper and lower bounds on the drag force experimental by a single bubble and/or swarms of bubbles. The applicability of this method is limited only to creeping flow situations involving the so-called generalized Newtonian fluids. Furthermore, strictly speaking this method yields true upper and lower bounds only in the case of Newtonian and power law liquids. Notwithstanding this fact, the results obtained by Jarzebski and Malinowski [1987], Gummalam et al. [1988], etc. for the Carreau viscosity equation must be treated with caution. The investigation of Nakano and Tien [1968], Mohan [1974], Chhabra and Dhingra [1986], etc. exemplify this class of solutions.

(ii)   The second type of solution is based on the so-called standard perturbation approach. This method involves expanding the solution in terms of a series in non-Newtonian material parameter, e.g. Weissenberg number. This scheme has often been used, along with viscoelastic constitutive equations. One would expect this type of solution to be applicable only for a small degree of deviation in fluid behaviour from the standard Newtonian case. Within the range of validity, only a very small effect of non-Newtonian flow behaviour is predicted. The contributions of Ajayi [1975] and Moo-Young and Hirose [1972] are representative of this approach.

(iii) Finally, the third class of solutions relies on the linearization of the governing equations, and hence for this approach also, the corresponding Newtonian result is used as the base solution, to which the additional effects arising from the generalized Newtonian flow behaviour are added. This approach has been extensively used by Hirose and Moo-Young [1969], and more recently by Bhavaraju et al. [1978], Kawase and Ulbrecht [1981], Kawase and Moo-Young [1985], etc. Intuitively one would expect the solutions of this type also to be applicable to only mildly non-Newtonian fluids.

In almost all of the aforementioned studies, the bubbles have been assumed to be spherical and of constant size, and free from surface active agents. Unfortunately, the wide variety of shapes of bubbles (as discussed in section 2) observed in non-Newtonian media not only renders such idealized analyses of little practical value but also precludes the possibility of making detailed comparisons between theory and experiments. Such comparisons are further hampered by the fact that the surface tension effects are not taken into account in the theoretical treatments, whereas the experimental work in this area is always influenced by the presence of naturally occurring impurities (which sometime act as surface active agents) even when all possible care is taken. Lastly, the experimental results also entail a degree of uncertainty owing to the wall effects and inadequate rheological characterization of the test fluids, etc. Figure 7 shows a typical comparison between theory and experiments for power law liquids under creeping flow conditions. The results are expressed in the form of a drag correction factor (X) defined as follows:

$$X = \frac{C_D Re}{24} \qquad (10)$$

where Re is the Reynolds number $\left(\frac{\rho U^{2-n} d^n}{K} \right)$ and C_D is the usual drag coefficient.

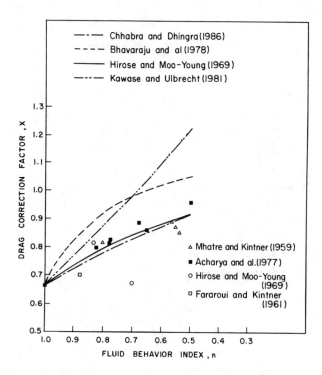

FIGURE 7.   Comparison between theory and experimental data for gas
            bubbles moving in power law liquids.

Admittedly, some investigators [Acharya et al., 1977; Yamanaka and
Mitsuishi, 1977], on the other hand, have resorted to experiments
and have proposed empirical formulae which relate drag coefficient to
Reynolds number and to other dimensionless parameters. None of
these, however, have been tested rigorously and are too tentative and
restrictive to be included here. Virtually nothing is known about the
influence of viscoelasticity, and Reynolds number on the behaviour of
bubbles.

It is readily agreed that one often encounters swarms of bubbles
rather than an isolated bubble. Recently, Gummalam et al. [1987,
1988] have employed the variational principles to obtain upper and
lower bounds on the rise velocity of a swarm of monosize spherical
bubbles rising through a quiescent body of power law and Carreau
model liquids. Their results for power law liquids are shown in Figure
8.

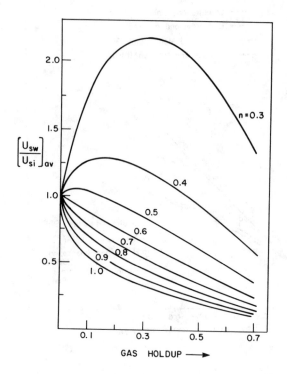

FIGURE 8. Swarm velocity for bubbles rising in a power law liquid
(replotted from Gummalam and Chhabra, 1987).

COALESCENCE

In process applications involving dispersions of gases into liquids,
bubbles are constantly colliding with each other. Depending upon the
bubble volumes and upon the velocity of collisions, bubbles may
separate from each other or coalesce. Owing to its overwhelming
pragmatic significance, considerable effort has been expended in
developing an improved understanding of the phenomenon of bubble
coalescence with the ultimate objective of integrating it with design
practices, especially in biotechnological applications.

It is generally agreed upon that the process of bubble coalescence
involves three steps: (i) initial contact between bubbles. This is
primarily controlled by the hydrodynamics of the bulk liquid, resulting
in a film of the order of a few μm, separating the two bubbles; (ii) a
gradual and continuous thinning of this film, to a thickness of a few
Angstroms. This process of film drainage is determined by the

58

TABLE II. Summary of investigations concerning the motion of bubbles in non-Newtonian media.

Investigator	Type of study	Fluid model	Details of work	Observations
Acharya et al. [1977, 1978]	Experimental	Power law	Air bubbles in aqueous solutions of carboxymethyl cellulose, polyacrylamide and polyethylene oxide	Results on bubble shapes, volume-velocity, drag coefficient and coalescence
Ajayı [1975]	Theoretical	Oldroyd fluid	-	Expressions for drag coefficient and shapes of bubbles
Astarita [1966]	Theoretical	Maxwell model	-	Qualitative results on the effect of fluid elasticity
Astarita and Appuzzo [1965]	Experimental	Power law	Bubbles in aqueous solutions of carbopol, carboxymethyl cellulose and ET-497	Velocity-size data and shapes of bubbles
Barnett et al. [1966]	Experimental	Power law and Ellis model	Aqueous solutions of carboxymethyl cellulose	Data on shapes, drag coefficient and Sherwood number
Bhavaraju et al. [1978]	Theoretical	Power law and Bingham plastics	-	Approximate expressions for drag coefficient and Sherwood number for single and swarms of bubbles

59

TABLE II. (Continued)

Bisgaard and Hassager [1982]	Experimental	-	Air bubbles in aqueous solution of polyacrylamide	Qualitative information on flow patterns around a bubble
Calderbank et al. [1970]	Experimental	Power law	CO_2 bubbles in aqueous solutions of polyox	Data on shapes, rise velocity and mass transfer
Carreau et al. [1974]	Experimental	Carreau model	Air bubbles in aqueous polyacrylamide and carboxymethyl cellulose solutions	Velocity-size relationship for bubbles in viscoelastic fluids and dimensional analysis
Chhabra and Dhingra [1986]	Theoretical	Carreau model	-	Approximate upper and lower bounds on drag coefficient
Dang et al. [1972]	Theoretical	Power law	-	Study on mass transfer with a chemical reaction
De Kee and Chhabra [1988]	Experimental	-	Viscoelastic solutions of polyacrylamide	Photographs on shapes and coalescence in viscoelastic media
De Kee et al. [1986, 1990]	Experimental	De Kee fluid model	CO_2, N_2 and air bubbles in aqueous solutions of carboxymethyl cellulose, and polyacrylamide	Data on size-velocity, and coalescence
Gummalam et al. [1987, 1988]	Theoretical	Power law and Carreau model	-	Upper and lower bounds on rise velocity of swarms of bubbles

Reference	Type	Fluid model	System	Remarks
Hassager [1977, 1979]	Experimental	-	Air bubbles in aqueous solutions of polyacrylamide	Characteristics of flow patterns and wakes
Hirose and Moo-Young [1969]	Theoretical	Power law	-	Approximate expressions for drag coefficient and Sherwood number
Jarzebski and Malinowski [1986]	Theoretical	Power law	-	Upper and lower bounds on drag coefficient and Sherwood number for swarms of bubbles
Kawase and Ulbrecht [1981 a, b]	Theoretical	Power law	-	Qualitative explanation of the discontinuity in size-velocity data
Kawase and Moo-Young [1985]	Theoretical	Carreau and Ellis fluid models	-	Approximate expressions for drag coefficient
Leal et al. [1971]	Experimental and semi-theoretical	No specific fluid model	Bubbles in aqueous solutions of Separan AP-30	Plausible explanation for discontinuity in size-velocity data
Macedo and Yang [1974]	Experimental	Power law	Air bubbles in aqueous solutions of Separan AP-30	No discontinuity in size-velocity data
Mitsuishi et al. [1972]	Experimental	Sutterby fluid model	Air bubbles in aqueous solutions of carboxymethyl cellulose and polyethylene oxide	Dimensionless correlation for drag coefficient

TABLE II. (Continued)

Mohan [1974], and Mohan and Raghuraman [1976]	Theoretical	Power law and Ellis models	-	Drag coefficient in creeping and intermediate Reynolds number regime
Mohan and Venkateswarlu [1974, 1976 a, b]	Theoretical	Ellis fluid model	-	Upper and lower bounds on drag coefficient
Nakano and Tien [1968, 1970]	Theoretical	Power law	-	Drag coefficient in creeping and intermediate range of Reynolds number region
Philippoff [1937]	Experimental	-	Air bubbles rising through time dependent fluids	Qualitative information on bubble shapes
Tiefenbruck and Leal [1982]	Theoretical	Oldroyd fluid model	-	Detailed flow field around a moving bubble
Yamanaka and Mitsuishi [1977]	Experimental	Sutterby fluid model	Air bubbles moving in aqueous solutions of polyethylene oxide and carboxymethyl cellulose	Empirical correlation for drag coefficient
Zana and Leal [1978]	Experimental and theoretical	Oldroyd fluid model	CO_2 gas bubbles in aqueous solutions of Separan AP-30	Effect of viscoelasticity on mass transfer

hydrodynamic forces prevailing in the film; (iii) the instant rupture of the film. Evidently, the rate of film drainage and thinning in step (ii) determines whether coalescence will be achieved. If the time required to thin the film to a few Ångstroms is longer than the period of contact between the two bubbles, coalescence will not occur. The last step (iii) is very rapid in comparison to the first two steps.

Utilizing these ideas, numerous models [Chesters and Hoffman, 1982; Dimitrov and Ivanov, 1978; Marrucci, 1969; Sagert and Quinn, 1978; Oolman and Blanch, 1986, etc.] have been proposed in the literature. However, most of these deal with Newtonian fluids, and entail varying degrees of approximations. In addition, the presented cases are restricted to highly idealized conditions such as two bubbles growing on adjacent orifices, involving bubbles in a prefixed geometrical con-figuration, etc. Hence, these are not directly applicable to real situa-tions. However, such studies have provided useful insights into the physics of the process and are also in *qualitative* agreement with experimental observations in this field.

In contrast, very little is known about the process of bubble coales-cence in rheologically complex systems. Additional complications arise when dealing with a viscoelastic continuous phase. Acharya and Ulbrecht [1978] have merely reported some preliminary results on the coalescence of in line bubbles and no attempt was made to explain the influence of rheological properties on coalescence. De Kee et al. [1986, 1990] have presented experimental results of a more complete study on the coalescence of bubbles, especially with regard to the initial geometrical arrangement of bubbles, the influence of fluid viscoelasticity, and the effect of surfactants. Furthermore, De Kee et al. [1990] used three different gases (air, nitrogen and carbon dioxide), to investigate the dimagnetic the effects. In this work, bubbles were simultaneously released from two orifices whose geometrical orientation could be varied by choosing the appropriate pair of orifices. One bubble was always of a fixed volume (V_f = 1, 2 or 3 x 10^{-6} m^3) whereas the volume of the second bubble (V_v) was measured in order to achieve coalescence (with the constant volume bubble) at pre-chosen heights (H) above the orifices.

The orifice separation is denoted by (S). Typical coalescence data are plotted in a dimensionless semi-log form in Figure 9. A cursury look at such a figure reveals a discontinuity in the plots of (V_v/V_f) vs (S/H) for a critical value of separation between the orifices.

Intuitively, it appears that in order to achieve coalescence for a large orifice separation, the required volume of the leading bubble (V_v) becomes quite large and this effect possibly appears as a discontinuity in the (V_v/V_f) vs (S/H) plots.

Figure 10 shows a photographic representation of two and three bubble coalescence. The picture is worth a thousand words, when it comes to describe the events leading to coalescence. The sequence of events leading to coalescence in the three bubble case can be postulated as follows: the central bubble positions itself ahead of the other two bubbles (presumably due to the nearby presence of the

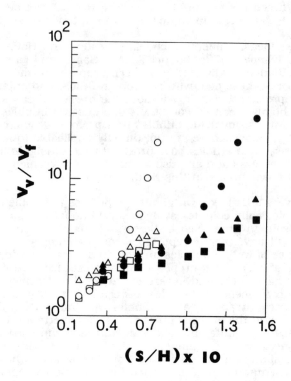

$(S/H) \times 10$

FIGURE 9. Effect of changing the coalescence height (H) and the volume of the fixed bubble (V_f) on the coalescence process of air bubbles in a 1.0 wt % solution of polyacrylamide in a 50/50 wt % mixture of water and glycerine. For H=0.25 m: ($\bullet$)V_f=1x10^{-6} m^3, ($\blacktriangle$) V_f=2x10^{-6} m^3 and ($\blacksquare$) V_f=3x10^{-6} m^3. For H=0.50 m: (O) V_f=1x10^{-6} m^3, ($\triangle$) V_f=2x10^{-6} m^3 and ($\square$) V_f=3x10^{-6} m^3.

other bubbles and vessel walls) and thus provides an opportunity for the trailing bubbles to enter its wake. It is unlikely that the wake behind the leading bubble would be large enough to accommodate both trailing bubbles simultaneously. In this particular instance, the bubble on the right undergoes a tremendous deformation into a banana shape and enters the wake of the leading bubble. Subsequently, the two bubbles collide and become separated by a thin film of liquid which eventually drains, ruptures and results in coalescence. A similar sequence of events leads to the coalescence with the third bubble.

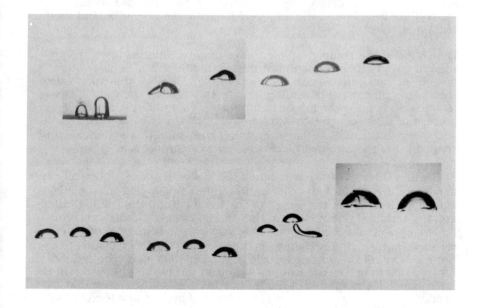

FIGURE 10. Bubble coalescence. Top row: same solution as in Figure 9. $V_1 = 9.33 \times 10^{-6}$ m^3 and $V_2 = 3.53 \times 10^{-6}$ m^3. Initial horizontal separation (S) between the two bubbles: 0.0024 m. Bottom row: coalescence in a 1.0 wt % aqueous solution of carboxymethyl-cellulose. $V_1 = V_2 = V_3 = 7.5 \times 10^{-6}$ m^3. Initial horizontal separation between bubbles: 0.003 m.

An examination of a plot such as Figure 9, reveals that the size of the leading bubble (V_v) decreases with increasing H. To compensate for the shorter time available in the case of small values of H, a larger wake capable of entrapping the trailing bubble is required.

Figure 9 also elucidates the effect of the volume of the trailing bubble ($V_f = 1, 2$ and 3×10^{-6} m^3). Evidently, a large trailing bubble requires a large leading bubble (hence, bigger wake) to achieve coalescence.

The viscoelasticity plays a significant role in the coalescence process by altering the shape and wake characteristics of the bubbles. Furthermore, orifice separation appears to be a significant parameter in determining whether coalescence will occur or not.

Our work (De Kee et al. 1990) showed that for low values of S and for different values of V_f and H, the smallest leading bubble volume required for coalescence occured in the case of highly viscoelastic solutions. This is simply because of the low terminal rise velocity of the trailing bubble and the relatively longer wake region behind the leading bubble.

The shape of the trailing bubble under these conditions is oblate-spheroid, and it is obvious that if the leading bubble also falls within this shape range, no coalescence is possible. Besides, this shape also shows irregular vortex shedding, and in this region, the wake characteristics do not appear to exert an appreciable influence on the process of coalescence (Narayanan et al., 1974). It is, however, necessary that the V_v be larger and that transition to a spherical-cap shaped bubble is attained before coalescence can occur.

For large values of orifice separation (S), a stable shape (spherical-cap) is adopted by the large leading bubble volume (V_v). The elastic effects enter the process via the length of the wake behind the leading bubble and the terminal rise velocity of the trailing bubble. Coalescence may no longer occur in the case of highly viscoelastic solutions, due to the sensitive nature of the relationship between $<V_t>$ and V. That is to say: if the volume of the leading bubble is smaller than that required for coalescence to occur, the wake fails to trap the trailing bubble. On the other hand, if the volume of the leading bubble is larger than that needed for coalescence to occur, the leading bubble accelerates too fast, leaving the trailing bubble behind. Eventually, coalescence could occur, but at a different height above the orifice plate.

Our results (De Kee et al. 1990) clearly showed also the complex roles played by the viscoelasticity, coalescence height (H) and bubble characteristics, including terminal velocity, wake and shape. For instance, in the case of large V_f values (3×10^{-6} m^3) and highly viscoelastic solutions, the coalescence process produces a transition in the V_v/V_f vs S/H plots at values of S, lower than would be the case for less elastic solutions.

Also, the leading bubble in a viscoelastic solution of reduced surface tension, required a slightly larger volume for coalescence to occur than that in a solution without added surfactant. One effect of the addition of surfactant is to suppress the internal circulation and this in turn somehow retards the process of film drainage and final rupturing. Therefore, a larger volume of the leading bubble (V_v) is required, i.e., the two coalescing bubbles will be subjected to high tangential surface stresses which in turn will promote internal circulation,thus, initiating the surface motion and speeding up the film thinning process.

Finally, we observed the need for larger V_v values, in order to achieve coalescence, when nitrogen was used instead of air. In the case of highly elastic solutions, effects of different gases seem to be suppressed by the high levels of the elastic forces.

Bailar et al. (1978) postulated that the different diamagnetic characteristics of the N_2 molecule results in a repulsion between the two bubbles and thus the rate of film drainage is reduced. Likewise, the larger volume of a leading CO_2 bubble needed for the coalescence to occur could be explained by invoking the degree of polarizability of CO_2 molecules. However, both these effects - diamagnetic and polarizability - are active at relatively small orifice separations and for small bubbles, and these effects do not seem to play any role at the larger orifice separations.

It is evident that the overall process of coalescence is governed by a balance between hydrodynamic, surface tension, elastic and electrostatic forces. The nature of these interactions is not at all clear at this stage. However, it is believed that the experimental results on coalescence presented herein would steer future theoretical developments in this area.

CONCLUSIONS

In this chapter, we have outlined the formation and the motion of bubbles in rheologically complex media. In particular, attention has been paid to the formation of single bubbles from submerged orifices in stagnant and in flowing liquids. Bubble formation and bubble shapes in non-Newtonian and in viscoelastic solutions has been discussed. Drag coefficient-Reynolds number relationships are also presented. The controversial terminal rise velocity-bubble volume plot has been examined using different gases. The cases of single and multiple bubbles in stationary liquids have been treated. A table highlighting a multitude of investigations has been presented. The effects of aging of polymer solutions and the presence of a surface active agent has also been examined in our laboratory. We emphasize, that we never observed a discontinuity in the velocity-volume curve. Qualitatively different shapes were observed in the case of the non-Newtonian viscoelastic liquids.

We also report on the coalescence of bubbles. The coalescence of two bubbles was examined under the conditions where one of the bubbles had a fixed volume. The volume of the second bubble required to realize coalescence was measured as a function of the height above the orifices, the separation between the two orifices, the type of gas, the presence of a surfactant and the aging of a test liquid. The influence of each of these variables has been analyzed, and possible physical arguments have been proposed which appear to explain the observed phenomena. It is, however, not yet possible to describe the process of coalescence analytically, but it is hoped that the experimental results presented will assist in formulating the future models for coalescence in viscoelastic media.

NOMENCLATURE

C_D	drag coefficient (-)
g	acceleration due to gravity (ms^{-2})
H	height above orifice at which coalescence was studied (m)
K	power law consistency index $(Pa.s^n)$
n	power law index (-)
Q	gas flow rate $(m^3 s^{-1})$
r_f	final radius of bubble (m)
R_c	critical size of bubble, eqn. 9 (m)
Re	Reynolds number (-)
S	distance between orifices (m)
t	time (s)
t_c	time of detachment (s)
U_{sw}	terminal rise velocity of a bubble swarm $(m\ s^{-1})$
U_{si} or $<V_t>$	terminal rise velocity of a single bubble under steady state conditions $(m\ s^{-1})$
V	volume of a bubble (m^3)
V_f	final volume of the bubble/volume of the fixed volume bubble in coalescence studies (m^3)
V_{fb}	volume of the force balance bubble (m^3)
V_v	volume of the variable volume bubble in coalescence (m^3)
x	distance (m)
X	drag correction factor $(= C_D Re/24)$ (-)
μ	viscosity of liquid phase (Pa.s)
ρ_L, ρ_G	density of liquid and gas respectively
σ	surface tension (Nm)
$\Delta\rho$	density difference $(= \rho_L - \rho_G)$ $(kg\ m^{-3})$

REFERENCES

Acharya, A. and Ulbrecht, J.: Note on the influence of viscoelasticity on the coalescence rate of bubbles and drops, A.I.Ch.E.J., 24, 348-351 [1978].

Acharya, A., Mashelkar, R.A. and Ulbrecht, J.: Mechanics of bubble motion and deformation in non-Newtonian media, Chem. Eng. Sci., 32, 863-872 [1977].

Acharya, A., Mashelkar, R.A. and Ulbrecht, J.: Bubble formation in non-Newtonian liquids, Ind. Eng. Chem. Fundam., 17, 230-232 [1978].

Ajayi, O.O.: Slow motion of a bubble in a viscoelastic fluid, J. Eng. Maths., 9, 273-280 [1975].

Astarita, G.: Spherical gas bubble motion through Maxwell liquids, Ind. Eng. Chem. Fundam., 5, 548-553 [1966].

Astarita, G. and Appuzzo, G.: Motion of gas bubbles in non-Newtonian liquids, A.I.Ch.E.J., 11, 815-820 [1965].

Bailar, J.C. et al.: General Chemistry, Academic Press, NY [1978].

Barnett, S.M., Humphrey, A.E. and Litt, M.: Bubble motion and mass transfer in non-Newtonian fluids, A.I.Ch.E.J., 12, 253-259 [1966].

Bhavaraju, S.M., Mashelkar, R.A. and Blanch, H.W.: Bubble motion and mass transfer in non-Newtonian fluids, A.I.Ch.E.J., 24, 1063-1076 [1978].

Bisgaard, C. and Hassager, O.: An experimental investigation of velocity fields around spheres and bubbles moving in non-Newtonian liquids, Rheol. Acta, 21, 537-539 [1982].

Blanch, H.W. and Bhavaraju, S.M.: Non-Newtonian fermentation broths: Rheology and mass transfer, Biotech. Bioeng., 18, 745-761 [1976].

Blass, E.: Formation and coalescence of bubbles and droplets, Int. Chem. Eng., 30, 206-211 [1990].

Bond, W.N. and Newton, D.A.: Bubbles, drops and Newton's law, Phil Mag., 5, 794-800 [1928].

Calderbank, P.H., Johnson, D.S.L., and Loudon, J.: Mechanics and mass transfer of single bubbles in free rise through some Newtonian and non-Newtonian liquids, Chem. Eng. Sci., 25, 235-256 [1970].

Carreau, P.J., et al.: Dynamique des bulles en milieu viscoélastique, Rheol. Acta, 13, 477-489 [1974].

Chesters, A.K. and Hoffman, G.: Bubble coalescence in pure liquids, App. Sci. Res., 38, 353-361 [1982].

Chhabra, R.P.: Hydrodynamics of bubbles and drops in rheologically complex liquids, Encyclopedia of Fluid Mech., 7, 253-286 [1988].

Chhabra, R.P. and Dhingra, S.C.: Creeping motion of a Carreau fluid past a Newtonian fluid sphere, Can. J. Chem. Eng., 64, 897-905 [1986].

Clift, R., Grace, J.R. and Weber, M.E.: Bubbles, drops and particles, Academic Press, NY [1978].

Costes, J. and Alran, C.: Models for the formation of gas bubbles at a single submerged orifice in a non-Newtonian fluid, Int. J. Multiphase Flow, 4, 535-551 [1978].

Dajan, A., M.A.Sc. thesis, University of Windsor [1985].

Dang, V., Gill, W.N., and Ruckenstein, E.: Unsteady mass transfer between bubbles and non-Newtonian liquids (Power law model) with chemical reactions, Can. J. Chem. Eng., 50, 300-306 [1972].

Davidson, J.F. and Harrison, D.: Fluidized particles, Cambridge Uni. Press, Cambridge [1963].

Davidson, J.F. and Schuler, B.O.G.: Bubble formation at an orifice in an inviscid liquid, Trans. Inst. Chem. Engrs., 38, 335-342 [1960].

De Kee, D., Carreau, P.J., and Mordarski, J.: Bubble velocity and coalescence in viscoelastic liquids, Chem. Eng., Sci., 41, 2273-2283 [1986].

De Kee, D. and Chhabra, R.P.: A photographic study of shapes of bubbles and coalescence in non-Newtonian polymer solutions, Rheol. Acta, 27, 656-660 [1988].

De Kee, D., Chhabra, R.P. and Dajan, A.: Motion and coalescence of gas bubbles in non-Newtonian polymer solutions, J. Non-Newt. Fluid Mech., in press [1990].

Dimitrov, D.S. and Ivanov, I.B.: Hydrodynamics of thin liquid films. On the rate of thinning of microscopic films with deformable interfaces, J. Colloid Interface Sci., 64, 97-106 [1978].

Fararoui, A. and Kintner, R.C.: Flow and shape of drops in non-Newtonian fluids, Trans. Soc. Rheo., 5, 369-380 [1961].

Ghosh, A.K. and Ulbrecht, J.: Bubble formation from a sparger in polymer solutions-I. Stagnant liquid, Chem. Eng. Sci., 44, 957-968 [1989]; ibid 969-977.

Gummalam, S., and Chhabra, R.P.: Rising velocity of a swarm of spherical bubbles in a power law non-Newtonian liquid, Can. J. Chem. Eng., 65, 1004-1008 [1987].

Gummalam, S., Narayan, K.A., and Chhabra, R.P.: Rise velocity of a swarm of spherical bubbles through a non-Newtonian fluid: Effect of zero shear viscosity, Int. J. Multiphase Flow, 14, 361-373 [1988].

Hassager, O.: Bubble motion in structurally complex fluids, Chem. Eng. with Per Soltoft, Teknisk for iagas, Kobenhavn, 105-117 [1977].

Hassager, O.: Negative wake behind bubbles in non-Newtonian liquids, Nature, 279, 402-403 [1979].

Hetsroni, G., Handbook of multiphase systems, McGraw Hill, NY [1982].

Hirose, T., and Moo-Young, M.: Bubble drag and mass transfer in non-Newtonian fluids: Creeping flow with power law fluids, Can. J. Chem. Eng., 47, 265-267 [1969].

Jarzebski, A.B., and Malinowski, J.J.: Transient heat and mass transfer from drops or bubbles in slow non-Newtonian flows, Chem. Eng. Sci., 41, 2575-2578; ibid. 2569-2573 [1986].

Jarzebski, A.B., and Malinowski, J.J.: Drag and mass transfer in creeping flow of a Carreau fluid over drops or bubbles, Can. J. Chem. Eng., 65, 680-684 [1987].

Kawase, Y., and Moo-Young, M.: Approximate solutions for drag coefficient of bubbles moving in shearthinning elastic fluids, Rheo. Acta, 24, 202-206 [1985].

Kawase, Y., and Ulbrecht, J.: On the abrupt change of velocity of bubbles rising in non-Newtonian liquids, J. Non-Newt. Fluid Mech., 8, 203-212 [1981a].

Kawase, Y., and Ulbrecht, J.: Newtonian fluid sphere with rigid or mobile interface in a shearthinning liquid: Drag and mass transfer, Chem. Eng. Commun., 8, 213-231 [1981b].

Kelkar, B.G., and Shah, Y.T.: Gas hold up and backmixing in bubble column with polymer solutions, A.I.Ch.E.J., 31, 700-702 [1985].

Kemblowski, Z., and Kristiansen, B.: Rheometry of fermentation liquids, Biotech. Bioeng., 28, 1474-1483 [1986].

Kezios, P. and Schowalter, W.: Rapid growth and collapse of single bubbles in polymer solutions undergoing shear, Phys. Fluids, 29, 3172-3181 [1986].

Krevelen, D.W.V. and Hoftiyzer, P.J.: Studies on gas bubble formation, Chem. Eng. Prog., 46, 29-35 [1950].

Kumar, R. and Kuloor, N.R.: The formation of bubbles and drops, Adv. Chem. Eng., 8, 255-368 [1970].

Leal, L.G., Skoog, J. and Acrivos, A.: On the motion of gas bubbles in a viscoelastic liquid, Can. J. Chem. Eng., 49, 569-573 [1971].

Macedo, I.C. and Yang, W.J.: The drag of air bubbles rising in non-Newtonian liquids, Jap. J. App. Phys., 13, 529-533 [1974].

Marrucci, G.: A theory of coalescence, Chem. Eng. Sci., 24, 975-985 [1969].

Mashelkar, R.A.: Rheological problems in chemical engineering, Rheology, 1, 219-233, Plenum Press, NY [1980].

Mhatre, M.V., and Kintner, R.C.: Fall of liquid drops through pseudoplastic liquids, Ind. Eng. Chem., 51, 865-867 [1959].

Mitsuishi, N., Yamanaka, A., and Sueyasu, Y.: Drag force on a moving bubble and droplet in viscoelastic fluids, Proc. PACHEC, 316-317 [1972].

Miyahara, T., Wang, W.-H. and Takahashi, T.: Bubble formation at a submerged orifice in non-Newtonian and highly viscous Newtonian liquids, J. Chem. Eng. Japan, 21, 620-626 [1988].

Mohan, V.: Creeping flow of a power law fluid over a Newtonian fluid sphere, A.I.Ch.E.J., 20, 180-182 [1974].

Mohan, V., and Raghuraman, J.: Viscous flow of an Ellis fluid past a Newtonian fluid sphere, Can. J. Chem. Eng., 54, 228-234 [1976].

Mohan, V., and Venkateswarlu, D.: Creeping flow of a power flow fluid past a fluid sphere, Int. J. Multiphase Flow, 2, 563-570 [1976].

Mohan, V., and Venkateswarlu, D.: Creeping flow of an Ellis fluid past a Newtonian fluid sphere, Int. J. Multiphase Flow, 2, 571-579 [1976].

Mohan, V., and Venkateswarlu, D.: Lower bound on the drag offered to a Newtonian fluid sphere placed in a flowing Ellis fluid, J. Chem. Eng. Japan, 7, 243-247 [1974].

Moo-Young, M., and Hirose, T.: Bubble mass transfer in creeping flow of viscoelastic fluids, Can. J. Chem. Eng., 50, 128-130 [1972].

Nakano, Y., and Tien, C.: Creeping flow of power law fluid over Newtonian fluid sphere, A.I.Ch.E.J., 14, 145-151 [1968].

Nakano, Y., and Tien, C.: Viscous incompressible non-Newtonian flow around fluid sphere at intermediate Reynolds number, A.I.Ch.E.J., 16, 569-574 [1970].

Narayanan, S., Goossens, L.H.J. and Kossen, N.W.F.: Coalescence of two bubbles rising in line at low Reynolds number, Chem. Eng. Sci., 29, 2071-2082 [1982].

Oolman, T. and Blanch, H.: Bubble coalescence in stagnant liquids, Chem. Eng. Commun., 43, 237-261 [1986].

Philippoff, W.: The viscosity characteristics of rubber solutions, Rubber Chem. Tech., 10, 76-104 [1937].

Rabiger, N., and Vogelpohl, A.: Bubble formation and its movement in Newtonian and non-Newtonian liquids, Encyclopedia of Fluid Mech., 3, 58-88 [1986].

Sagert, N.H. and Quinn, M.J.: The coalescence of gas bubbles in dilute aqueous solutions, Chem. Eng. Sci., 33, 1087-1095 [1978].

Subramaniyan, V. and Kumar, R.: Tech. Rep. #CE/67-5, Indian Institute of Science, Bangalore [1967].

Tiefenbruck, G., and Leal, L.G.: A numerical study of the motion of a viscoelastic fluid past rigid spheres and spherical bubbles, J. Non-Newt. Fluid Mech., 10, 115-155 [1982].

Tsuge, H. and Terasaka, K.: Volume of bubbles formed from an orifice submerged in highly viscous Newtonian and non-Newtonian liquids, J. Chem. Eng. Japan, 22, 418-421 [1989].

Yamanaka, A., and Mitsuishi, N.: Drag coefficient of a moving bubble and droplet in viscoelastic fluids, J. Chem. Eng. Japan, 10, 370-374 [1977].

Zana, E., and Leal, L.G.: The dynamics and dissolution of gas bubbles in a viscoelastic fluid, Int. J. Multiphase Flow, 4, 237-262 [1978].

Chapter 3

The Motion of a Sphere Through an Elastic Fluid

K. WALTERS
Department of Mathematics
University College of Wales
Aberystwyth, UK

R. I. TANNER
Department of Mechanical Engineering
University of Sydney, Australia

1. INTRODUCTION

The motion of a rigid sphere through a Newtonian viscous fluid at low Reynolds number is one of the oldest problems in theoretical fluid mechanics. The problem dates from the pioneering work of Stokes [1] who developed an analytic solution for the case of vanishingly small Reynolds numbers and an unbounded fluid. Specifically, Stokes showed that the drag F on a stationary sphere in a uniform stream of constant magnitude U is given by

$$F = 6\pi\eta Ua, \tag{1}$$

where a is the radius of the sphere and η the viscosity. Numerous attempts have been made to generalize (1) to include the effects of bounding walls, non-zero Reynolds numbers and non-spherical obstacles.

Non-Newtonian effects have also received considerable attention in recent years and the basic problem is now viewed as being of fundamental importance in non-Newtonian fluid mechanics. The reasons for this are:

(i)   the experiment can be performed in the laboratory with *relative* ease;

(ii)   the basic flow is non-viscometric and the symmetrical flow of a sphere through a cylindrical container provides a bench-mark problem for practitioners in the numerical simulation of viscoelastic flow [see, for example, 2, 3]. Of particular attraction is the fact that the problem does not have any geometrical singularities.

It is probably true to say that enough experimental and theoretical work on the non-Newtonian problem has been carried out for a reasonably clear picture to be drawn concerning the distinctive general effects of non-Newtonian rheology. However, there are still gaps in our knowledge of the experimental effects of viscoelasticity on the drag on a sphere and available numerical simulation studies are able to provide little more than a qualitative picture of observed behaviour.

Non-Newtonian elastic liquids have many characteristics which are not present in the case of Newtonian liquids. Accordingly it is often difficult to estimate which particular non-Newtonian effect is having the strongest influence on flow characteristics. For example, for most non-Newtonian liquids, the viscosity in a viscometric flow varies with the shear rate and variable viscosity effects must play a role in the non-viscometric sphere problem. We would also expect viscoelasticity *per se* to influence the drag on the sphere and it is often impossible to distinguish between these influences. One way forward is to employ recent

73

simulations [see, for example 4, 5] for *in*elastic non-Newtonian fluid models which cater for variable viscosity effects, in order to remove that particular influence on the drag and so concentrate specifically on the effect of viscoelasticity. An alternative and more popular procedure has been to carry out experiments on the so called Boger [6] fluids, which have an (almost) constant viscosity and yet are highly elastic. In this way, any departure from Stokes' formula (1) can be unambiguously associated with viscoelasticity. Such a procedure has the immediate advantage that many numerical algorithms for constant-viscosity fluid models are already available and hence comparison between experimental observation and numerical simulations is facilitated.

In the present chapter, we give some attention to the problem of a sphere moving in a general elastic liquid; but our main concern is with the experimental data which is available for constant-viscosity Boger fluids and with the growing number of numerical simulations which are available for comparison. Interestingly, a seemingly simple experiment often turns out to be anything but simple, and, in the case of numerical simulation, the advantage of a lack of a geometrical singularity is sometimes countered by the presence of regions of strong stress concentration which present formidable challenges to numerical analysts.

If we now concentrate on the problem of a sphere of radius a falling through a cylindrical tube of radius R, we may define the important geometrical ratio a|R as ß. Further, if λ is a characteristic time of the elastic liquid (obtained for example from the first normal-stress difference and shear stress in a viscometric flow [7]) we may define a non-dimensional elasticity number W_e (sometimes called the Weissenberg number and at other times the Deborah number) through

$$W_e = \lambda \frac{U}{a} . \tag{2}$$

The Reynolds number R_e is defined by

$$R_e = \frac{\rho U a}{\eta} , \tag{3}$$

where ρ is the fluid density and (for a Boger fluid) η is the (constant) viscosity.

For the steady-state drag F on the sphere, we may now write, in place of (1),

$$F = 6\pi\eta U a K, \tag{4}$$

where

$$K = K(\beta, R_e, W_e). \tag{5}$$

In the following, we shall not discuss the dependence of K on the Reynolds number in great detail. Rather, we shall concentrate on the effect of ß and especially W_e. In Fig 1 we show the effect of ß on K for a Newtonian liquid (K_N) (i.e. $W_e = 0$).

2. EXPERIMENT

2.1 General

A survey of the theoretical and experimental literature would suggest that in general terms a drag reduction is evident for small W_e (i.e. $K < K_N$) whilst the opposite may be the case for large W_e (i.e. $K > K_N$). This is an example of the general behaviour predicted by Barnes &

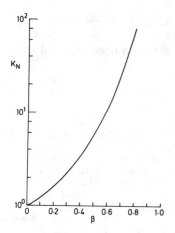

FIGURE 1. Wall correction factor K as a function of ß for a rigid sphere moving axially through a Newtonian liquid in a cylindrical container [8]. Note that, for ß > 0.5, the results in [8] are often insufficiently accurate, see text.

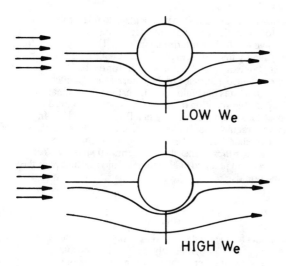

LOW W_e

HIGH W_e

FIGURE 2. Schematic diagram of the shift in streamlines for low and high W_e. Note that in Newtonian flow at zero Reynolds number (Stokes flow) the streamlines are symmetric.

Walters [9]. Also, for small W_e, streamlines are shifted slightly downstream while an upstream shift is associated with high values of W_e (as shown in Fig 2). Some of these conclusions will be discussed in more detail later, experimentally for Boger fluids and numerically for constant-viscosity rheological models.

2.2 The 'Time' Effect

Experiments by Bisgaard [10] and Cho et al [11] have demonstrated one unexpected experimental difficulty with the seemingly-simple falling-ball experiment. These authors showed that, for highly elastic polymer solutions, the terminal velocity can be strongly dependent on the time interval between the dropping of successive balls. For example, Bisgaard concludes that if spheres are released each 10th minute, the terminal velocity of the spheres can be up to 30% higher than the first sphere released in the liquid. Even with an hour between each dropping there can still be a 9% increase in velocity. Cho et al speculate that, when a ball drops through a highly elastic polymer solution, it opens or breaks the high molecular weight polymer network matrix locally. As the ball moves down the centreline of the cylinder the solvent immediately fills the space left by the ball in the centre region. Therefore the fluid in the centre region has a relatively higher concentration of solvent, thus making spheres fall faster than in the undisturbed liquid.

The scale of the time-interval effect is certainly such as to shout a warning to any experimenters making falling-ball measurements.

2.3 Overshoot

A popular and simple method of performing falling-ball experiments is to release the balls from rest and time their descent by, for example, a stroboscopic technique. It is then possible to ascertain when the balls have reached their terminal velocity U. Very often, for Newtonian liquids, the time (or equivalently the distance) required for balls to reach their terminal velocity can be very modest. For example, it is not unusual for the balls to reach their terminal velocity after they have travelled a distance of the order of their diameter. Experiments indicate that for Newtonian liquids the build up from rest to the terminal velocity is monotonic. For elastic liquids the situation can be much more complex. Figure 3 contains a picture of successive positions of a sphere falling in a Boger fluid [12]. In this case, a distance of nearly 20 diameters is required before the sphere reaches its terminal velocity. Also evident is the sizable overshoot. Here, there is certainly no monotonic behaviour and the maximum velocity value is *three* times the final terminal velocity. Yet again, it would be dangerous to take accepted practice in Newtonian fluid mechanics as the norm in experiments on highly elastic liquids.

2.4 Wall Effects

Most of the available experimental results for the elastic-liquid flow problem involve essentially an infinite expanse of test liquid (i.e. ß=0) and only limited data are available for moderate values of ß. What little there is might point to the following conclusions:

(i)   For a fixed value of ß < 0.15, the effect of wall proximity (at least for constant viscosity elastic liquids) can be accommodated by using the Newtonian data shown in Fig 1[see for example 13, 14]. However, for shear thinning inelastic fluids, the effect of the walls certainly diminishes as the non-Newtonian effects increase. For example, in the family of power law fluids ($\sigma = k\dot{\gamma}^n$) as n → 0, the effect of the walls vanishes rapidly [cf. 5].

(ii)   For values of ß in the range 0.3 to 0.5, Bisgaard [10] found that the observed

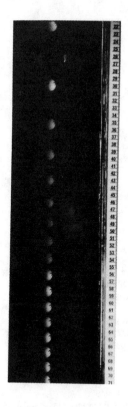 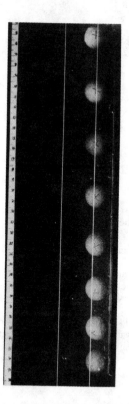

Fig 3. Successive positions of a sphere falling in a Boger fluid (after Walters 12]. The sphere is of radius 1.6cm and was released at the 20cm mark on the scale.

Fig 4. The movement of a falling sphere towards a plane wall in the case of a 2% aqueous solution of polyacrylamide. The vertical lines are drawn to illustrate the sideways motion of the sphere towards the right hand wall. Note the rotation of the sphere in this experiment as demonstrated by the successive positions of the cross on the sphere [12]

decrease in K for a 1% solution of polyacrylamide in glycerol was more steep for high values of ß than for low ones [see also 15, 16].

More experiments on the dependency of K on ß for non zero W_e are required before a definitive general picture can be drawn. Such experiments must be carried out with care to ensure that the spheres fall along the axis of the cylinder, without rotation [cf. 44, 45]. Walters [12] has shown that, when spheres fall from an initially asymmetrical position in the cylinder, viscoelasticity results in a strong force at right angles to the motion and directed *towards* the cylinder container. Figure 4 illustrates the effect for spheres falling close to a plane wall.

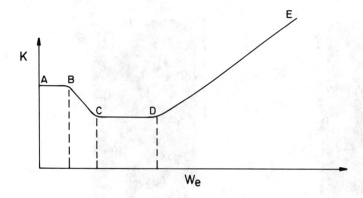

FIGURE 5. Schematic diagram of experimental (K, W_e) data for Boger fluids.

2.5   Other General Studies

High Reynolds number studies on the flow of inelastic and elastic liquids past a sphere were carried out by Acharya et al [17, 18]. In the process they highlighted the possibility of delayed boundary-layer separation and the formation of a 'dual-wake'.

Riddle et al [19] studied the distance between *two* identical spheres falling along their line of centres in an elastic liquid. They found that, for small initial separations of the spheres, the two spheres eventually converge. In contrast, for large initial separations, the two spheres eventually diverge.

2.6   Experiments on Constant-Viscosity Boger Fluids

There is a growing literature on the fall of spheres through constant-viscosity elastic liquids [12-14, 16-20] and, although the general picture is provocative and often confusing, it is possible to postulate the behaviour shown schematically in Fig 5, at least for the case ß=0.

The horizontal portion of the graph between A and B is to be expected from continuum mechanics requirements (cf § 3). Its presence has been confirmed by Mena et al [16], Tirtaatmadja et al [13] Chhabra et al [20] and Chmielewski et al [14] up to $W_e \sim 0.1$. These authors also observed the drag reduction implied by the portion of the curve from B to C with Mena et al and Chhabra et al demonstrating a fall of approximately 25% from the Newtonian value at $W_e = 0$, the point C being located at $W_e \sim 1.0$. The results of Tirtaatmadja et al and Chmielewski et al are more equivocal, the latter authors finding the same trend for a corn-syrup based Boger fluid (with $K \sim 0.9$ at the highest available experimental value of W_e (~0.2)) but with virtually no drag reduction for polybutene-based Boger fluids. Some of their experimental data are consistent with a drag reduction of 5%, but this is certainly within experimental error. The experiments of Tirtaatmadja et al on the polybutene-based test liquid designated M1 point to a drag reduction of at most 10%. The difference in response between corn-syrup based and polybutene based Boger fluids in the experiments of Chmielewski et al demands further study.

The experiments of Mena et al are consistent with the existence of a plateau region CD for values of W_e between 1 and 2 (the highest available value). The oft quoted data of Chhabra et al have also been used to support the existence of an asymptotic value of K, but a detailed study of their data might alternatively support a turning point, with no perceptible CD region. What is clear is that when results at relatively high W_e values are available, there is

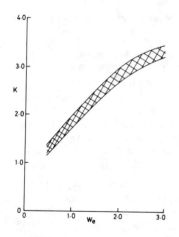

FIGURE 6. The shaded area represents the spread of data for a series of maltose-syrup based Boger fluids, ß=0 [12].

quite a substantial increase in drag. For example, the data of Tirtaatmadja et al show a K value of approximately 1.3 at the highest available W_e value (~2) with every indication of further drag enhancement at higher values of W_e. The results of Walters [12] are even more dramatic (see Fig 6) with values of K of approximately 3 being obtained at a W_e of 3, this time for a series of maltose-syrup based Boger fluids.

It is clear from the foregoing discussion that the overall trend suggested in §2.1 of drag reduction at low W_e followed by drag enhancement at higher W_e is not inconsistent with the majority of available experimental data on constant-viscosity Boger fluids. It is also clear that further experiments are required to resolve the difference in behaviour of seemingly similar elastic liquids. Hopefully, such experiments will also elucidate the finer details of the behaviour postulated in Fig 5.

3.   THEORY

3.1   General

The research activity in the *experimental* falling-ball problem has been matched by a growing *theoretical* literature. The early papers written in the 60's were necessarily of an analytical nature and were essentially perturbation problems with the Weissenberg number W_e as the perturbation parameter [see, for example, 21-23]. Later, Mena and Caswell [24] included inertia terms in formally matched asymptotic expansions for small but non-zero R_e. These early analyses predicted a small change in the streamline pattern from the purely viscous case with a shift in the downstream direction in agreement with experiments.

The observed *upstream* shift in the streamlines for high values of W_e has been predicted by Ultman and Denn [25]. The relevant analysis has been criticized by Zana et al [26], but it is still regarded as an important contribution since it indicated for the first time the possibility of a 'change of type' in the governing equations as W_e increases beyond some critical value.

It is unlikely that a change of type can occur at zero Reynolds number and/or in the presence of a solvent viscosity.

So far as the drag on the sphere is concerned, the plateau region AB is implied by all the low W_e analyses and the departure from the Stokes drag is predicted to be a quadratic function of W_e. In qualitative terms the analyses are consistent with the drag reduction implied by the portion BC of the curve in Fig 5. In order to study this region in detail and also investigate the possibility of predicting drag enhancement at higher values of W_e, it is really necessary to consider full numerical solutions of the governing equations. However, Chilcott and Rallison [27] [see also 28, 29] have investigated the flow around an unbounded sphere using a dumbbell-type fluid model. The dumbbells are of finite length L; as $L \rightarrow \infty$ one reverts to the Oldroyd B model, but for finite L, the extensional viscosity remains bounded. In the limit $W_e \rightarrow 0$, the Newtonian result is obtained. They showed that a narrow wake where stresses are high develops behind the sphere and great care was needed to resolve the wake numerically. The results indicate that the drag coefficient K reaches a constant value as $W_e \rightarrow \infty$ for the cases when L is finite. Initially K falls as W_e increases for both large and small L. As W_e increases beyond unity, K has a minimum at around $W_e=1$ and then rises to an asymptotic greater than 1 for L=10; for L=2.5 the drag falls monotonically to an asymptotic value less than 1. We believe that the papers [27-29] provide a good representation of the physics of a real dilute viscoelastic fluid. (In the limit $L \rightarrow \infty$, where one recovers the Oldroyd/Maxwell fluid, the results are not as clear.) Generally, the picture shown in Fig 5 is consistent with the predictions of references 27-29.

Up to the present time viscoelastic *numerical* simulations have usually concentrated on constant-viscosity continuum models and are therefore relevant to the discussion of §2.6 on experiments for Boger fluids.

3.2 Numerical simulations for constant-viscosity viscoelastic models

One of the benchmark problems in the frequently-held International Numerical Simulation Workshops [see, for example, 2, 3] concerns viscoelastic flow past a sphere in a cylindrical tube, with ß = 0.5. Most of the simulations have been carried out for the so called Upper Convected Maxwell model with constitutive equations [cf. 30, 31]

$$\sigma_{ik} = -p\delta_{ik} + T_{ik}, \qquad (6)$$

$$T_{ik} + \lambda \overset{\triangledown}{T}_{ik} = 2\eta d_{ik}, \qquad (7)$$

where σ_{ik} is the stress tensor, p an isotropic pressure and d_{ik} is the rate of strain tensor.

∇ denotes the upper convected time derivative of Oldroyd [32, 33], η is a constant viscosity and λ is a relaxation time. Equations (6) and (7) are solved in conjunction with the familiar equations of motion and continuity, which for flows assumed to be isothermal, creeping and incompressible may be expressed in the form

$$\frac{\partial \sigma_{ik}}{\partial x_k} = 0, \qquad (8)$$

$$\frac{\partial v_k}{\partial x_k} = 0, \tag{9}$$

where v_k is the velocity vector and usual tensor notation is implied.

There are several estimates of the Newtonian drag force in a tube [see 8]. The Bohlin formula, obtained by the method of reflections [8] gives

$$K = [1.0 - 2.10444\beta + 2.08877\beta^3 - 0.94813\beta^5 - 1.372\beta^6 + 3.87\beta^8 - 4.19\beta^{10}]^{-1}. \tag{10}$$

For $\beta=0.5$, the last two terms in (10) are of size 0.0151 and 0.0041, respectively, so the error in the estimate of K is likely to be order 0.01. Formula (10) gives the value K=5.923 for $\beta=0.5$. Other 'exact' results of Haberman [see 8] give, for $\beta=0.5$, the value K=5.970. Numerical results of Lunsmann et al [41] and Zheng et al [42] depend on mesh sizes and are given in Table1 for $\beta=0.5$; these are the most accurate sets of values in the literature; some other values are also given.

The extrapolated value is expected to be within 0.005% of the true value; the Bohlin formula (10) and the Haberman 'exact' result are thus in error by about 0.4% at $\beta=0.5$ and not accurate enough to use as a standard (for $\beta=0.25$ by contrast, the Bohlin formula is certainly within about 0.01% of the true value). We shall take K=5.947 as the Newtonian standard for $\beta=0.5$. Numerical simulations of the viscoelastic flow problem have used several different methods [34-42]. Table 2 shows the investigations for the flow of a Maxwell fluid past a sphere. Most of the numerical studies have failed to obtain convergent results for values of W higher than 0.7, with few exceptions [37,40]. This is a manifestation of the so called 'high Weissenberg number problem'.

TABLE 1. Newtonian drag on a sphere in a tube ($\beta=0.5$). The polar mesh size refers to the grid or element size near the poles of the sphere.

Investigators	K	Degrees of freedom	Polar mesh size/a	cpu time	Machine
Hassager & Bisgaard [34]*	5.99	250	0.1	10 min	IBM 3033
Sugeng & Tanner [35]	5.92	1900(FEM)	0.5	~ 1 hr	Vax 11/780
Sugeng & Tanner [35]	5.92	100(BEM)	0.5	~ 20 min	Vax 11/780
Luo & Tanner [36]	5.98	~2000	0.25	~ 1 hr	Vax 11/780
Crochet [37]	5.98	-	0.1	-	-
Carew & Townsend [38]	6.00	1815	~ 1.0	1 min	Cyber 205
Lunsmann et al [41]	5.9477	19719	.0116	-	
	5.9476	40333	.0061	~ 3 min per iteration	Cray YMP
	5.9474	50010	.0020	-	
Zheng et al [42]	5.9423	72	.0436	-	
	5.9463	180	.0175	-	
	5.9466	240	.0131	~ 1 hr	MIPS 120/5
Extrapolated value	5.947	-	-		

*Value for $W_e = 0.011$ - method not applicable for $W_e = 0$.

TABLE 2. Summary of numerical work on the flow of a Maxwell fluid past a sphere in a cylinder; ß=0.5. FEM stands for Finite Element Method and BEM for the Boundary Element Method. Consult original papers for other definitions.

Investigators	Method	Limit on W_e
Hassager & Bisgaard [34]	Lagrangian FEM	0.57
Sugeng & Tanner [35]	BEM	0.7
Sugeng & Tanner [35]	FEM	0.3
Luo & Tanner [36]	Streamline FEM	0.4
Crochet [37]	Mixed FEM	-
Carew & Townsend [38]	FEM	0.6
Debbaut & Crochet [39]	FEM (with	No limit
Debbaut & Crochet [40]	upwinding)	No limit
Lunsmann et al [41]	FEM	~ 2
Zheng et al [42]	BEM	~ 0.6

Table 3 and Fig 7 contain numerical simulation data for the viscoelastic flow problem (ß = 0.5). There is clearly no problem of accuracy in the Newtonian case if sufficient elements are used. For $W_e \neq 0$, the results are diverse, especially at higher W_e and the breakdown at $W_e = 0(1)$ is noticeable for most methods. Large stress concentrations downstream from the rear stagnation point are now accepted and it is natural to wonder if a limiting W_e exists for steady axisymmetric flow [43].

TABLE 3. Viscoelastic Drag Results $K(W_e)$ for ß = 0.5.

Investigators	K(0)	K(0.2)	K(0.4)	K(0.6)	K(0.8)	K(1.0)
Hassager & Bisgaard [34]	5.99	5.38	4.81	4.27		
Sugeng & Tanner [35]	5.92	5.29	4.69	4.11		
Luo & Tanner [36]	5.98	5.10	4.82	-		
Crochet [37]	5.98	5.64	5.35	5.06	4.99	4.98
Carew & Townsend [38]	6.0	5.34	5.0	4.94		
Debbaut & Crochet [39]	5.96	5.62	5.21			
Armstrong et al [41]	5.95	5.65	5.20	4.80	4.53	4.36
Zheng et al [43] (Mesh M5)	5.92	5.6	5.00	4.5		

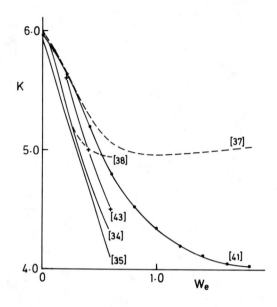

FIGURE 7. (K, W_e) numerical simulation data for $ß = 0.5$

4. CONCLUSIONS

The complete experimental picture concerning the drag on a sphere in a viscoelastic liquid is slowly emerging and there is little doubt that, for a suitably defined drag parameter K and elasticity parameter W_e, viscoelasticity often results in drag reduction for small W_e followed by drag enhancement for higher values of W_e. The initial drag reduction has been simulated qualitatively in all respectable computational and analytical work. However, there is little quantitative agreement, even for the same continuum model, and more work in this area is warranted. The ß=0 predictions of Rallison et al [27-29] and Crochet [37] as well as those of Debbaut and Crochet [39,40] for ß = 0.5 offer hope that the observed drag enhancement at moderate W_e is accessible to theoretical development, but further work in this area is also warranted.

Concerning the computational bench-mark problem discussed in the previous section, it *may* be that the particular value of ß = 0.5 was not a felicitous choice and ß=0.2 would have been better. The stress concentration near the trailing stagnation point is known to be one strong influence in determining the drag [27-29, 43], but when the gap between the sphere and the cylinder is small (ß → 1) another mechanism is also known to be operative, namely that associated with the converging and diverging flow regions between the sphere and the cylinder surfaces [9]. At some intermediate value of ß, the two mechanisms are likely to interact and, with present uncertainties, it may be beneficial to concentrate on the two extreme situations, namely ß=0 and ß → 1. This awaits further study.

NOMENCLATURE

a   Radius of sphere.

ß   Geometrical ratio (a/R).

d_{ik} Rate-of-strain tensor.

δ_{ik} Kronecker delta.

F   Drag force.

K   Drag factor (see equation 4).

K_N Drag factor for Newtonian liquid.

λ Relaxation time.

η Viscosity.

ρ Density.

R_e Reynolds number.

R   Radius of cylindrical container.

σ_{ik} Stress tensor.

T_{ik} Extra-stress tensor.

U   Velocity of sphere or velocity of stream.

v_k Velocity vector.

W_e Weissenberg number.

REFERENCES

1.   Stokes, G. C., Trans Camb Phil Soc 9, 8-106, 1851.

2.   Fourth workshop on numerical methods in viscoelastic flow. J non-Newtonian Fluid Mechanics 20, 1986.

3.   Fifth workshop on numerical methods in viscoelastic flow. J non-Newtonian Fluid Mechanics 29, 1988.

4.   Bush, M. B. and Phan-Thien, N., J non-Newtonian Fluid Mechanics 16, 303-313, 1984.

5.   Gu, D. and Tanner, R. I., J non-Newtonian Fluid Mechanics 17, 1-12, 1985.

6.   Boger, D. V., J non-Newtonian Fluid Mechanics 3, 87-91, 1977.

7.   Binding, D. M., Walters, K., Dheur, J. and Crochet, M. J., Phil Trans Roy Soc A323, 449-469, 1987.

8.   Happel, J. and Brenner, H., *Low Reynolds number hydrodynamics*, Noordhoff, Leiden, 1973.

9.   Walters, K. and Barnes, H. A., *Rheology Vol 1 Principles* ed G Astarita et al, 45-62 Plenum Press, 1980.

10.   Bisgaard, C., J non-Newtonian Fluid Mechanics 12, 283-302, 1983.

11.   Cho, Y. I., Hartnett, J. P. and Lee, W. Y., J non-Newtonian Fluid Mechanics 15, 61-74, 1985.

12.   Walters, K., To be published.

13.   Tirtaatmadja, Uhlherr, P. H. T. and Sridhar, T., J non-Newtonian Fluid Mechanics, to appear 1990.

14.   Chmielewski, C. Nichols, K. L. and Jayaraman, K., J non-Newtonian Fluid Mechanics, to appear 1990.

15.   Sigli, D. and Coutanceau, M., J non-Newtonian Fluid Mechanics 2, 1-21, 1977.

16.   Mena, B., Manero, O. and Leal, L. G., J non-Newtonian Fluid Mechanics 26, 1987, 247-275.

17.   Acharya, A., Mashelkar, R. A. and Ulbrecht, J., Rheologica Acta 15, 454-470, 1976.

18.   Acharya, A., Mashelkar, R. A. and Ulbrecht, J., Rheologica Acta 15, 471-778, 1976.

19.   Riddle, M. J., Narvaez, C. and Bird, R. B., J non-Newtonian Fluid Mechanics 2, 23-35, 1977.

20.   Chhabra, R. P., Uhlherr, P. H. T. and Boger, D. V., J non-Newtonian Fluid Mechanics 6, 187-199, 1980.

21.   Leslie, F. M. and Tanner, R. I., Quart J Mech Appl Math 14, 36-48, 1961.

22. Caswell, B. and Schwarz, W. H., J Fluid Mechanics 13, 417-426, 1962.

23. Giesekus, H. Rheologica Acta 3, 59-71, 1963.

24. Mena, B. and Caswell, B., Chem Eng J, 8, 125-134, 1974.

25. Ultman, J. S. and Denn, M. M., Chem Eng J, 2, 81-89, 1971.

26. Zana, E., Tiefenbruck, G. and Leal, L. G., Rheologica Acta 14, 891-898, 1975.

27. Chilcott, M. D. and Rallison, J. M., J non-Newtonian Fluid Mechanics 29, 381-432, 1988.

28. Harlen, O. G., Rallison, J. M. and Chilcott, M. D., J non-Newtonian Fluid Mechanics 34, 319-349, 1990.

29. Harlen, O. G., J non-Newtonian Fluid Mechanics. To appear.

30. Tanner, R. I., *Engineering Rheology* Revised Edition, Oxford University Press, 1988.

31. Barnes, H. A., Hutton, J. F. and Walters, K., *An Introduction to Rheology*, Elsevier, 1989.

32. Oldroyd, J. G., Proc Roy Soc A200, 523-541, 1950.

33. Oldroyd, J. G., J non-Newtonian Fluid Mechanics 14, 9-46, 1984.

34. Hassager, O. and Bisgaard, C., J non-Newtonian Fluid Mechanics 12, 153-164, 1983.

35. Sugeng, F. and Tanner, R. I., J non-Newtonian Fluid Mechanics 20, 281-292, 1986.

36. Luo, X-L and Tanner, R. I., J non-Newtonian Fluid Mechanics 21, 179-199, 1986.

37. Crochet, M. J. Proc Xth Int Congress on Rheology, Sydney, Paper 1.19, 1988.

38. Carew, E. and Townsend, P., Rheologica Acta 27, 125-129, 1988.

39. Debbaut, B. and Crochet, M. J., J non-Newtonian Fluid Mechanics 30 169-184, 1988.

40. Debbaut, B. and Crochet, M. J., Lecture presented at 6th Numerical Simulation workshop in Hindsgavl, 1989.

41. Lunsmann, W. J., Brown, R. A. and Armstrong, R. C., J Sci Computing. To appear 1990.

42. Zheng, R., Phan-Thien, N. and Tanner, R. I., J non-Newtonian Fluid Mechanics. To appear 1990.

43. Phan-Thien, N., J non-Newtonian Fluid Mechanics. To appear 1990.

44. Christopherson, D. G. and Dowson, D., Proc Roy Soc A251, 550-564, 1959.

45. Tanner, R. I., Chem Eng Sci 19, 349-355, 1964.

Chapter 4

The Effect of Surfactants on Flow and Mass Transport to Drops and Bubbles

G. C. QUINTANA
AT&T Bell Laboratories
600 Mountain Avenue
Murray Hill, New Jersey 07974

1. INTRODUCTION

The hydrodynamic motion of fluid drops in an unbounded continuous phase fluid is hindered significantly by the presence of surface adsorbing contaminants. The physical mechanism underlying this retardation was first described by Frumkin and Levich (1947), (as reviewed by Levich, 1962). When a droplet translates through a continuous liquid phase which contains soluble surface-active contaminant, the surfactant diffuses towards the drop, adsorbs onto the interface, and is subsequently convected to the trailing pole. Accumulation of molecules at the rear, leads to a surface diffusive flux towards the front stagnation point, and to local desorption of surfactant to the adjacent bulk fluid sublayer. The total concentration of adsorbed surfactant reaches a steady state value which reflects the balance of these transport processes (Figure 1). Owing to the local elevation in the surfactant concentration at the drop rear, an interfacial tension force arises, which opposes the surface flow. It is this Marangoni stress which reduces the drop motion.

In describing the hydrodynamics of a surfactant covered drop, the interfacial tension, σ, enters through the force balance at the drop surface. In the absence of surface viscosity, the shear stress difference at the surface is proportional to the interfacial tension gradient, while the jump in the normal stress is equal to 2σ times the mean surface curvature, H. Since the interfacial tension varies with Γ, the concentration of surfactant, solution of the governing fluid mechanics equations requires the specification of an equation of state

$$\sigma = \sigma(\Gamma), \tag{1}$$

and simultaneous determination of the local surfactant distribution.

The rate of surface convection relative to that of the other transport processes, provides two bounds on the surfactant gradient. If surface convection is much slower than surface diffusion, adsorption-desorption, or bulk diffusion, then local accumulation of surfactant at the rear pole is minimal. In this limit, the surfactant concentration around the drop varies little from Γ_0, the concentration value which is realized under zero-flow conditions. Because of the small local variation in the interfacial tension, the Marangoni stress uniformly retards the surface flow, and the drop translational velocity is reduced only slightly from the Hadamard-Rybczynski, or clean surface values. Alternately, if the rate of surface flow is rapid relative to the other transport mechanisms, then

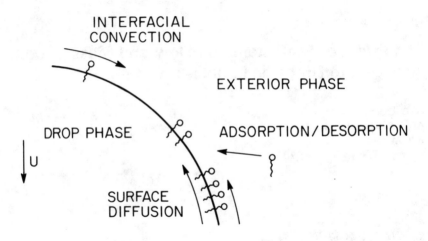

INTERFACIAL
CONVECTION

EXTERIOR PHASE

DROP PHASE

ADSORPTION/DESORPTION

U

SURFACE
DIFFUSION

Figure 1:   Transport process at the drop interface.

surfactant molecules will be swept to a cap region at the drop rear. In this case, the gradient in Γ is steep, and the resulting Marangoni tension is large enough to suppress all surface mobility in the "stagnant cap" region.

Because of the strong coupling of the field equations, the surface mass conservation equation, and the stress boundary conditions, exact solutions for this class of problems are impossible, except in the limits of "uniform retardation" (Levich, 1962) and "stagnant cap" (Sadhal and Johnson, 1983) flow (Figure 2). Although much of the literature in this area has been restricted to the retarded flow regime, the distribution of surfactant for systems of practical importance will in fact be skewed toward the drop rear. Numerical simulations (Holbrook and Levan, 1983; Quintana, 1986) of the governing equations are required for the general hydrodynamic problem. In this paper, we discuss the hydrodynamic and transport results for three problems in which a variable interfacial tension gradient exists due to presence of surface active contaminants. In the following section, a general formulation for the contaminated Stokes flow problem is given in which the dimensionless parameters characterizing the various transport processes are defined. The effect of adsorbed surfactant on dropwise solute mass transport in the stagnant cap flow limit is examined in section 3. In section 4, the effect of continuous phase fluid elasticity and shear thinning viscosity on the terminal velocity attained by a surfactant laden Newtonian fluid droplet is described. Finally, we examine the influence of strong repulsive interactions between adsorbed contaminant molecules on the translational terminal velocity in the fifth section. We do this by relaxing the generally assumed linear relationship between the interfacial tension and the surfactant concentration, replacing this by the more realistic Frumkin constitutive relation.

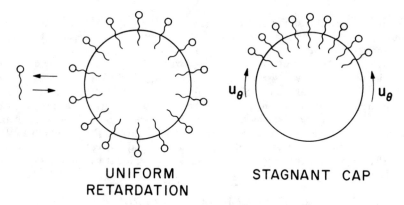

UNIFORM
RETARDATION

STAGNANT CAP

Figure 2: Surfactant distribution for uniform retardation and stagnant cap flows.

2. GENERAL FORMULATION

Normalization of the field equations and of the boundary conditions yields nondimensional parameters which characterize the relative rates of the transport processes. In the limit of small Reynolds and small capillary numbers, i.e. $\rho Ua/\mu \ll 1$ and $\mu U/\sigma \ll 1$, inertial effects are negligible and the interfacial tension is large enough to prevent viscous deformation of the droplet. (Note the U is the characteristic drop terminal velocity, a is the drop radius, μ and ρ are the continuous phase fluid viscosity and density, respectively.) The governing flow equations can be formulated in terms of axisymmetric spherical stream functions. Because of the assumption that the drop remains spherical, satisfaction of the local normal stress boundary condition is precluded (Levan and Newman, 1976) and this condition can be replaced by a macroscopic force balance from which the translational terminal velocity is determined. The tangential stress difference across the spherical interface is given in dimensionless form by

$$\tau_{r\theta} - \hat{\tau}_{r\theta} = \text{Ma} \frac{\partial \Gamma}{\partial \theta}. \tag{2}$$

where the interfacial tension is assumed to vary linearly with Γ. We define the Marangoni number as $\text{Ma} \equiv RT\Gamma_0/(\mu U)$, i.e. characterizing the ratio of surface pressure force to the force due to viscous shear. For Marangoni number vanishingly small, the interfacial force exerted is negligible and drop translational motion is unaffected by the presence of surfactants. As Ma approaches infinity, the strength of the interfacial tension force completely immobilizes the interface and the drop settles as a solid sphere.

If the bulk Peclet number is large, then diffusion of contaminant to the interface occurs rapidly and the distribution of surfactant will be determined solely by surface convection, surface diffusion and sorption. Scaling the surface mass conservation equation by U and Γ_0 yields

$$\frac{\partial \Gamma}{\partial t} + \frac{1}{\sin\theta} \frac{\partial}{\partial \theta}(v_s(\theta)\Gamma\sin\theta) = \text{Pe}_s^{-1}(\frac{1}{\sin\theta} \frac{\partial}{\partial \theta}(\sin\theta \frac{\partial \Gamma}{\partial \theta})) - \text{Bi}\,(\Gamma - 1). \tag{3}$$

The surface Peclet number, $Pe_s \equiv Ua/D_s$, is typically on the order of 10^5 because of small surface diffusivity. The Biot number as defined by ha/U, characterizes the ratio of the sorption rate, h (s^{-1}), to the rate of surface convection. In the limit of infinitely large Bi, the rate of transfer of contaminant between the surface and the boundary layer surrounding the drop is rapid. This fast exchange of molecules, precludes large Γ gradients, and consequently the drop mobility is unaffected. For smaller Biot number, the sorption process is slower; the confinement of surfactant to the surface and surface convection couple to give rise to large gradients in Γ and to substantial reduction of the translational velocity. Stagnant cap behavior results in the limits $Pe_s \rightarrow \infty$ and $Bi \rightarrow 0$ where the surface distribution is determined soley by convection.

3. MASS TRANSPORT TO A STAGNANT CAP DROP

The adsorption of surfactants at fluid interfaces is known to hinder evaporation and mass transport. The effect of surfactant adsorption on evaporation at stagnant planar fluid interfaces has been investigated by Rideal (1925), Langmuir and Schaefer (1943), and by Archer and La Mer (1955). These authors attempted to correlate the steric hindrance caused by the adsorption of surfactants (i.e. as manifested by the extent of evaporation hindrance) with monolayer thickness and alkyl chain length. The extension of this work to systems in which the fluid interface is in motion, date back to experimental studies on dropwise solute mass transfer in the presence of surfactants by Garner and Hale (1953) and Lindland and Terjesen (1956). In these studies, the magnitude of interfacial solute transport to falling drops was reduced by up 67% relative to transport in clean systems. The large order of magnitude of these reductions were confirmed by subsequent investigations by Huang and Kintner (1969) and by Mekasut, Molinier and Angelino (1978). Although terminal velocity reductions are reported in all of these studies, only Huang and Kintner observed the interfacial flow. These authors reported a vortex shift and a streamline asymmetry which are consistent with the existence of a stagnant cap region, as had been previously reported by Savic (1953), Garner and Skelland (1955), Elzinga and Banchero (1961), Griffith (1962), and Horton et.al. (1965) in other systems.

For the stagnant cap drop, solute mass transfer is complicated by the fact that the adsorbed monolayer has a twofold effect on interfacial mass transport. First, surfactant retards interfacial hydrodynamic motion and thereby reduces convective transport; and secondly, the adsorbed monolayer presents a steric hindrance to the passage of solute molecules analogous to that which exists for stagnant interfacial evaporation. The extent of interfacial transport impedance will depend on the nature of the surfactant film. A highly compressed film of bulk insoluble surfactant is capable of inhibiting interfacial transfer while bulk soluble surfactants produce no discernible barrier for the transport of small diffusing species (Plevan and Quinn (1966), Burnett and Himmelblau (1970), and Mudge and Heideger (1970)). Boundary layer analysis of surfactant effects on solute mass transport such as that presented by Lochiel (1965) is precluded in the case of the stagnant cap drop owing to both the nonuniformity of the surface mobility and the steric influence of the surfactant cap. In this analysis, we isolate the effect of steric hindrance from the hydrodynamic influence of surfactant adsorption by integrating the convective diffusion equation for dropwise mass transfer subject to two boundary conditions. We first assume that the stagnant cap completely blocks the passage of solute. We then relax this condition and solve the mass transport problem for a monolayer which offers no physical obstruction to solute transport. The boundary conditions which correspond to surface blocking are zero flux on the cap and finite solute concentration on the surfactant free leading portion of the drop. In the absence of steric hindrance, the reduced convection boundary conditions require a constant interfacial solute concentration over the entire drop surface. After briefly reviewing the stagnant cap hydrodynamics, we compare the mass transfer Sherwood numbers obtained in these two limits.

Stagnant cap flow occurs when desorption is kinetically limited and contaminant molecules are thus confined to the drop interface. In this limit the Biot number tends to zero, and the steady surface mass balance equation (3) reduces to

$$\nabla_s \cdot (v_s \Gamma) = 0, \tag{4}$$

which has the solution

$$v_s = 0 \qquad 0 \le \theta < \phi \tag{5}$$

in the cap region, and

$$\Gamma = 0 \qquad \phi \le \theta \le \pi \tag{6}$$

on the clean surface. Note that ϕ is the stagnant cap angle as measured from the rear stagnation point as shown in Figure 3 and its value depends on the total amount of adsorbed surfactant. The solution of the momentum equations subject to these restrictions on v_s and $\Gamma(\theta)$ must account for the discontinuity of the tangential velocity and the divergence of the tangential stress in the vicinity

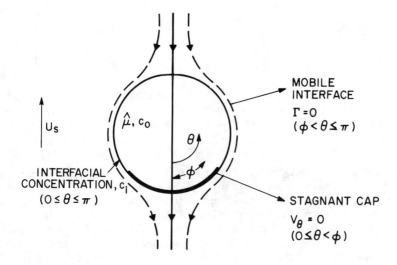

Figure 3: The stagnant cap drop.

of the cap edge. Because of this difficulty, convergence of the inner and outer stream functions requires a large number of radial coefficient terms (Savic (1953), Davis and Acrivos (1966) and Harper (1973, 1982)). Sadhal and Johnson's (1983) reformulation of the momentum equations and boundary conditions as a set of dual series equations, allowed determination of the stream function coefficients and of the translational terminal velocity as analytical functions of the cap angle ϕ. The stagnant cap terminal velocity is given by

$$U_s(\phi) = \left\{ \frac{(3\hat{\mu}+2\mu)U_H}{\dfrac{\mu}{2\pi}(2\phi+\sin\phi-\sin2\phi-\dfrac{1}{3}\sin3\phi)+2\mu+3\hat{\mu}} \right\} \tag{7}$$

where μ and $\hat{\mu}$ are the exterior and drop viscosities, and U_H is the Hadamard-Rybczynski terminal velocity.

We consider solute transport from a stagnant cap drop to the continuous phase. If the diffusivity and the initial concentration of the solute in the disperse phase are large, the drop phase concentration can be assumed to be constant. In this limit, determination of the exterior concentration field requires integration of the steady exterior convective diffusion equation

$$\left\{ \frac{\partial^2 C}{\partial r^2} + \frac{2}{r}\frac{\partial C}{\partial r} + \frac{1}{r^2\sin\theta}\frac{\partial}{\partial\theta}\sin\theta\frac{\partial C}{\partial\theta} \right\} = \frac{Pe}{2}\left\{ v_r(r,\theta)\frac{\partial C}{\partial r} + \frac{v_\theta(r,\theta)}{r}\frac{\partial C}{\partial\theta} \right\}. \tag{8}$$

Note that the stagnant cap velocities, $v_r(r,\theta)$ and $v_\theta(r,\theta)$ have been normalized by U_H, and the lengths by the drop radius, a. The dimensionless concentration is $C(r,\theta) = [c(r,\theta) - c_\infty]/[c_i - c_\infty]$, where c_∞ and c_i are the solute concentrations far away from the drop and in the sublayer adjacent to the drop. The exterior phase Peclet number is $Pe = 2U_H a/D$, where D is the solute diffusion coefficient. The interfacial boundary conditions which correspond to surface blocking are therefore

$$\frac{\partial C(1,\theta)}{\partial r} = 0 \qquad 0 \leq \theta < \phi, \tag{9}$$

and

$$C(1,\theta) = 1 \qquad \phi < \theta \leq \pi \tag{10}$$

while far away from the drop, the solute concentration drops to zero. In the absence of steric hindrance, the dimensionless concentration is unity for all θ. The governing mass transfer equation was integrated with the reduced convection and the surface blocking boundary conditions using the finite element method.

The presence of a stagnant cap of $\phi > 0.1\pi$, reduces solute mass transport significantly relative to transport in a clean system. The Sherwood numbers decrease with increasing cap angle and increasing Peclet number. As expected, the extent of this transfer hindrance is greater for the surface blocking cap than for the freely penetrable cap. An interesting finding from these simulations, is that the excess transport hindrance due to surface blocking does not scale with the percent surface coverage, and therefore the influence of physicochemical steric blocking cannot be deduced solely from geometric arguments. In the following paragraphs, we summarize these findings.

The calculated Sherwood numbers are normalized with respect to Sherwood number values obtained in the absence of surfactant. For the clean drop, the surface is freely mobile and the net transport, $Sh(\kappa, Pe, \phi=0)$, is maximum (where $\kappa=\hat{\mu}/\mu$). The normalized transfer ratios are plotted in Figure 4 as a function of Peclet number for $\kappa=1$ and different cap angles for the reduced convection limit. With increasing cap size, the Marangoni stress acting on the drop increases, and convection is thereby reduced. In the absence of steric hindrance, the decrease of Sh with ϕ stems entirely from this reduced convection. The Peclet number dependence can be understood as follows. As convection begins to dominate diffusion, i.e. $Pe>1$, solute transport is determined primarily by the hydrodynamics. In the presence of contaminants (i.e. fixed ϕ), the reduced interfacial velocity curtails transport via convection, an effect which becomes more acute for increasing Peclet number. In Figure 5, we compare this Peclet number dependence with that exhibited for the surface blocking case. The Sherwood numbers for $\kappa=1$ and $\phi=0.5\pi$ are plotted versus Peclet number for reduced convection and surface blocking boundary conditions. We note that the surface blocking Sherwood numbers are always lower than the corresponding reduced convection values, and that in addition, they exhibit a substantially weaker Peclet number dependence. This diminished sensitivity of the Sherwood numbers on the convective transport mechanism, reflects the fact that for systems in which the surfactant sterically obstructs the passage of solute molecules, the hindered flux largely determines the net interfacial transport independent of the local hydrodynamics.

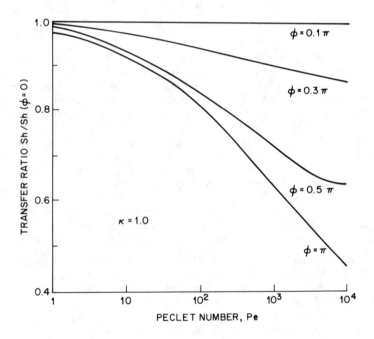

Figure 4:   Reduced convection normalized Sherwood numbers plotted as a function of Peclet number for different ϕ.

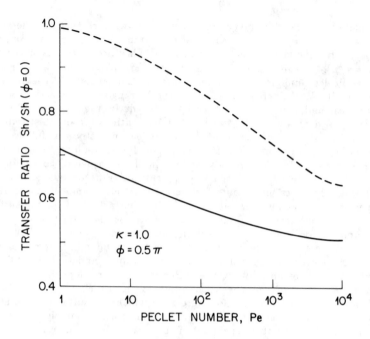

Figure 5: Surface blocking and reduced convection Sherwood numbers plotted versus the Peclet number for $\phi = 0.5\pi$.

A comparison of the surface blocking values with those calculated on the basis on stagnant cap hydrodynamics, show that the additional transfer reductions due to steric hindrance do not scale uniformly with ϕ. In Figure 6, we plot the absolute difference between Sherwood number values calculated under reduced convection and surface blocking conditions, as a function of the surfactant covered drop surface area. In this manner, we isolate the excess hindrance which stems directly from steric blocking from that due to Marangoni reduced convection. A plot of this normalized Sherwood number difference (Figure 6) for $Pe = 10^4$, suggests two distinct transport regimes. For $\phi > 0.4\pi$, the excess transport hindrance due to steric blockage scales geometrically, i.e. linearly with ϕ, such that in the limit of $\phi = \pi$, the adsorbed contaminant monolayer completely impedes transport. For the smaller coverages, the steric hindrance exhibits a weaker than ϕ dependence. In this low coverage regime, solute mass transport is independent of the steric influence until the coverage exceeds 20%.

Our simulation results show that steric blocking is unimportant for systems in which only trace amounts of contaminants are present, but may substantially hinder dropwise mass transport if a critical bulk concentration of surfactant is exceeded. For the stagnant cap drop, the steric effect is evident only when the cap covers more than 20% of the total interface. In this case, the full surface blocking boundary conditions predict the maximum retardation which can result from physicochemical steric hindrance. The magnitude of the steric blockage effect indicated by this upper limit analysis, points out the need for

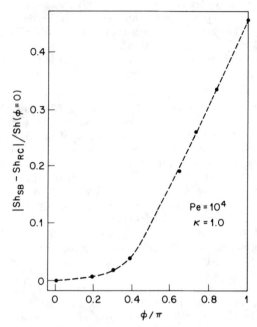

Figure 6: The absolute Sherwood number difference versus ϕ for Pe $= 10^4$.

considering the steric influence on mass transport for dropwise systems of practical importance for which only partial steric hindrance occurs.

4. THE EFFECT OF CONTINUOUS PHASE VISCOELASTICITY

Fluid particle motion in a viscoelastic fluid is complicated by the influence of shear thinning and increased extensional viscosities. If surface active species are present in the bulk or disperse phase, accumulation of adsorbed surfactant creates a surface pressure which opposes interfacial convection and reduces surface velocity. This surface velocity retardation increases the shear rate around the moving drop. When the continuous phase is Newtonian, the increase in shear rate leads simply to a increased drag and thus to a reduction in the drop terminal velocity. However, if the continuous phase is viscoelastic, several factors interact to determine the particle motion. First, when the surface velocity retardation is significant, shear thinning should act to decrease the drag acting on the particle. For a more mobile particle, the drag will be greatly influenced by the increased extensional viscosity which is characteristic of elastic fluids and which would serve to increase the extensional drag experienced by the drop. Finally, the secondary flow caused by the viscoelastic fluid acts to redistribute adsorbed surfactant and thus to increase the Marangoni stress. In this analysis, we resolve these competing effect by determining the motion of a surfactant laden fluid drop in a four constant Oldroyd fluid.

The basic understanding of the hydrodynamics of fluid particle motion in a viscoelastic phase has been established primarily through theoretical studies of retarded motion flow. These are translational flows in which the characteristic time for a fluid element to transverse the drop (a/U, where a and U are the drop radius and velocity) is much slower than the intrinsic viscoelastic relaxation time (denoted λ_1). The ratio of these scales defines the Weissenberg number, Wi $\equiv \lambda_1 / aU$. In the limit as $\lambda_1 / aU \rightarrow 0$, the fluid approaches near Newtonian behavior, and the fluid stresses are described by retarded motion expansions of the constitutive equation which are equivalent to n^{th} order fluids.

Retarded motion solutions for the steady translation of a solid sphere were studied by Leslie (1961), Caswell and Schwarz (1962), and Giesekus (1963). These studies demonstrated that the correction to the terminal velocity is second order in the Weissenberg number, a result of the fore-aft symmetry of the Newtonian flow. For the solid sphere, this second order correction is always positive, since shear thinning largely determines the sphere motion. Due to the shear thinning viscosity, the drag acting on the sphere in a viscoelastic fluid is less than that in a Newtonian one of viscosity equal to the zero shear rate viscosity. The solid sphere results contrast significantly with those of a gas bubble which were obtained by Tiefenbruck (1980). For the gas bubble, the viscoelastic correction to the Newtonian terminal velocity is again second order in Weissenberg number, but the velocity may be hindered or accelerated relative to the Hadamard-Rybczynski or Newtonian values. Tiefenbruck used a viscoelastic model in which the shear thinning and elongation viscosity characteristics could be varied such that their effect on the hydrodynamics could be ascertained. In the limit when the extensional viscosity is large, the enhanced elongational drag reduces the drop translational velocity and the bubble slows down. Alternately, when shear thinning dominates, the drag is reduced and the bubble accelerates relative to the Newtonian terminal velocity value. The solid sphere and the bubble behavior were confirmed by Quintana et.al. (1987) in an analogous study concerning droplets of finite viscosity.

4.1 The Viscoelastic Model

Oldroyd differential models are useful because they correctly describe the important features of fluid viscoelasticity. The four constant Oldroyd model with an intermediate objective time derivative was used to determine the effect of shear thinning and elongational viscosity on the translational motion of a surfactant laden Newtonian drop. The stress strain relationship is given by

$$\tau_j^i + \text{Wi} \left[\frac{D}{Dt} \tau_j^i - \alpha(\tau_m^i \dot{\gamma}_j^m + \dot{\gamma}_m^i \tau_j^m) \right] = 2 \left[\dot{\gamma}_j^i + \text{Wi} \cdot \epsilon \left(\frac{D}{Dt} \dot{\gamma}_j^i - 2\alpha(\dot{\gamma}_m^i \dot{\gamma}_j^m) \right) \right], \quad (11)$$

τ_j^i and γ_j^i are the ij^{th} components of the stress and strain rate tensors, which have been normalized by $\eta_0 U_0 / a$ and U_0 / a respectively. Note that η_0 is the zero shear rate viscosity. The Weissenberg number defined as $\lambda_1 U_0 / a$, where λ_1 is the relaxation time. The dimensionless parameter ϵ is the ratio of the retardation to relaxation times, λ_2 / λ_1. Variation of the fourth parameter α between zero and one permits a variable objective time derivative. When $\alpha = 0$, the time derivative (given by the expression in brackets) is the corotational or Jaumann type. When $\alpha = 1$, tensor differentiation is performed with reference to a co-deformational frame.

The most important unidirectional shear flow experiment is a steady simple shear. We can represent the dimensionless shear viscosity as a Taylor series expansion in the Weissenberg number

$$\eta/\eta_0 = 1 - \text{Wi}^2(1 - \alpha^2)(1 - \epsilon). \tag{12}$$

Note that for α equal to one, no shear thinning occurs, while for α near zero, maximum shear thinning is predicted. An analogous Taylor series expression can be derived for the elongational viscosity. Here we define the Weissenberg number to be $(= \lambda_1 \dot{\text{E}})$, where $\dot{\text{E}}$ is the characteristic rate of extension. Retaining the first non-zero correction term to the Newtonian extensional viscosity, we find

$$\bar{\eta}/3\eta_0 = 1 + \alpha\text{Wi}(1 - \epsilon) \tag{13}$$

Therefore as α increases, $\bar{\eta}$, the elongational viscosity increases.

The constitutive equations for the drop can be written in dimensionless form as

$$\hat{\tau}_{kj} = 2\kappa\hat{\dot{\gamma}}_{kj} \tag{14}$$

where the caret distinguishes the drop phase variables and κ is the ratio of the drop to the outer fluid zero shear rate viscosity. The viscoelastic pressure excluded stress can be expressed the sum of a Newtonian contribution and a group of nonlinear terms which arise from the Oldroyd constitutive expression

$$\underline{\tau} = 2\underline{\dot{\gamma}} + \underline{\Omega}(\underline{\tau}, \underline{\dot{\gamma}}, \kappa, \alpha, \epsilon, \text{Wi}). \tag{15}$$

4.2 Formulation

Substitution of the axisymmetric spherical stream functions $\hat{\psi}$ into the momentum equations leads to the fourth order homogeneous equation

$$\text{E}^4\hat{\psi} = 0, \tag{16}$$

for the drop phase, where

$$\text{E}^2 = \frac{\partial^2}{\partial r^2} + \frac{\sin\theta}{r^2}\frac{\partial}{\partial\theta}\left[\frac{1}{\sin\theta}\frac{\partial}{\partial\theta}\right]. \tag{17}$$

Owing to the nonlinearity of the Oldroyd stress strain constitutive model, the continuous stream function equation is inhomogeneous

$$\text{E}^4\psi = \text{Z}(r,\theta) \tag{18}$$

where $\text{Z}(r, \theta)$ is given in terms of the non-Newtonian stress contribution by

$$\text{Z}(r,\theta) = \sin\theta\left\{\frac{\partial}{\partial\theta}[\nabla\cdot\underline{\Omega}]_r - \frac{\partial}{\partial r}(r[\nabla\cdot\underline{\Omega}]_\theta)\right\}. \tag{19}$$

In the limit of weak elasticity and "slow flow", i.e. small Weissenberg number, a retarded motion flow analysis can be introduced such that the kinematic, stress and concentration variables can be expressed as Taylor series expansions in ascending powers of the Weissenberg number. The first non-zero viscoelastic correction to U_0, the terminal velocity attained by a surfactant laden drop in a Newtonian continuous phase fluid with viscosity equal to η_0,

$$U(\text{Wi}) = U_0 + \text{Wi}U^{(1)} + \text{Wi}^2U^{(2)} + \cdots, \tag{20}$$

is sought.

The kth order stream function equations are solved subject to the perturbation boundary conditions:

$$\frac{\partial \psi^{(k)}}{\partial r} = \frac{\partial \hat{\psi}^{(k)}}{\partial r} \quad \text{at } r = 1, \tag{21}$$

$$\tau_{(r\theta)}^{(k)} - \hat{\tau}_{(r\theta)}^{(k)} = \text{Ma} \frac{\partial \Gamma^{(k)}}{\partial \theta} \quad \text{at } r = 1, \tag{22}$$

$$\psi^{(k)} = \hat{\psi}^{(k)} = 0 \quad \text{at } r = 1, \tag{23}$$

$$\lim_{r \to 0} \hat{\psi}^{(k)}/r^2 \text{ is finite}, \tag{24}$$

$$\psi^{(k)} = \frac{1}{2} \sin^2 \theta \frac{U^{(k)}}{U_0} \quad \text{as } r \to \infty \tag{25}$$

Note that Ma is the Marangoni number ($= RT\Gamma_0 / (\eta_0 U_0)$) where Γ_0, is the equilibrium amount of adsorbed surfactant, and the surface tension has been assumed to depend linearly on the Γ. Due to the assumption that the drop does not deform, i.e. the capillary number, $\text{Ca} = (\eta_0 U_0) / \sigma$ is less than one, the interfacial normal stress balance is satisfied exactly. The hydrodynamic description is completed by satisfaction of the kth order macroscopic balance of the viscous forces

$$0 = 2\pi \int_0^\pi \left[(-p^{(k)} + \tau_{(r\theta)}^{(k)})|_{r=1} \cos\theta - \tau_{(r\theta)}^{(k)}|_{r=1} \sin\theta \right] \sin\theta \, d\theta \tag{26}$$

for $k \geq 1$.

The stream functions can be represented as infinite series products of r and θ (Happel and Brenner, 1965). For the viscoelastic fluid, the kth order stream function depends explicitly on the non-Newtonian contribution $Z(r,\theta)$

$$\psi^{(k)} = \sum_{n=2}^{\infty} [A_n^{(k)} r^n + \tag{27}$$

$$+ (r^{n+2}/(2(2n+1)(2n-1))) \int_1^r \varsigma^{-n+1} Z_n^{(k)}(\varsigma) \, d\varsigma$$

$$+ (r^n/(2(2n-1)(3-2n))) \int_1^r \varsigma^{-n+3} Z_n^{(k)}(\varsigma) \, d\varsigma$$

$$+ (r^{3-n}/(2(3-2n)(1-2n))) \int_1^r \varsigma^n Z_n^{(k)}(\varsigma) \, d\varsigma$$

$$+ (r^{-n+1}/(2(2n+1)(1-2n))) \int_1^r \varsigma^{2+n} Z_n^{(k)}(\varsigma) \, d\varsigma] C_n^{-1/2}(\cos\theta),$$

where $C_n^{-1/2}(\cos\theta)$ is the nth order Gegenbauer polynomials of degree -1/2. The hydrodynamic problem is solved by determining the inner and outer stream function constants in terms of $\Gamma^{(k)}$, subject to satisfaction of the boundary conditions (21) to (25). Note that $\Gamma^{(k)}$ is represented by the infinite Legendre polynomial expansion

$$\Gamma^{(k)}(\theta) = \sum_{n=0}^{\infty} a_n^{(k)} P_n(\cos\theta). \tag{28}$$

Numerical values for $a_n^{(k)}$ are determined from the kth order surface mass balance equations

$$\nabla_s \cdot (v_\theta{}^{(k-j)} \Gamma^{(j)} + v_\theta{}^{(j)} \Gamma^{(k-j)}) = \frac{1}{Pe_s} \nabla_s^2 \Gamma^{(k)} - Bi(\Gamma^{(k)} - \delta_{k0}) \qquad (29)$$

for $k \geq 1$, using the Galerkin weighted residual method (Finlayson, 1972) and the zero order surfactant coefficients as previously determined by Quintana, (1986). The kth order terminal velocity correction is obtained from the macroscopic force balance:

$$\frac{U^{(k)}}{U_0} = \frac{-3\kappa}{3\kappa + 2} \int_1^\infty r^{-2} \int_0^\pi \Omega_{(r\theta)}^{(k)}(r) C_2^{-1/2}(\cos\theta) \, d\theta \, dr \qquad (30)$$

$$+ \frac{1}{4(3\kappa + 2)} \int_1^\infty (3\kappa r^{-2} - (3\kappa + 2)) \int_0^\pi \left(\Omega_{(\theta\theta)}^{(k)} + \Omega_{(\phi\phi)}^{(k)} - 2\Omega_{(rr)}^{(k)} \right) \sin\theta P_1(\cos\theta) \, d\theta \, dr$$

$$+ 2 \frac{Ma}{3(3\kappa + 2)} a_1^{(k)}$$

4.3 Results and Discussion

The first non-zero correction is first order, $U^{(1)}/U_0$, and negative for all values of the viscoelastic parameters and the two dimensionless groups related to the properties of the surfactant, Bi and Ma. The influence of viscoelasticity is therefore to retard the drop motion relative to the value achieved in a Newtonian continuous phase. The major effects on the drag are contributed by shear thinning, increased extensional viscosity and redistribution of the surfactant due to secondary flows induced by viscoelasticity. Shear thinning tends to decrease the first order drag whereas the enhancement of the extensional viscosity increases it. In the Oldroyd model used, increasing α decreases the extent of shear thinning and simultaneously increases the extensional viscosity. We examine the effect of α on the drag.

Figure 7 illustrates the variation of $U^{(1)}/U_0$ on α for a gas bubble for a Marangoni number equal to one and for different values of the Biot number. For all values of the Biot number, as α increases the correction decreases linearly and hence the droplet velocity is further retarded. The $U^{(1)}/U_0$ dependence on the Biot number exhibits a sharp transition around Bi=1. For a fixed α and Marangoni number, as the Biot number increase from zero, the correction steadily decreases for values of Bi up to one. For Bi larger than one, the dependency reverses itself and $U^{(1)}/U_0$ tends towards zero with rising Bi (Figure 8). The Marangoni dependence of the first order correction is illustrated by comparing Figures 7 and 9. In Figure 9, $U^{(1)}/U_0$ is plotted against α for different Bi where Ma=10. Note that with an increase in Marangoni number from 1 to 10, $U^{(1)}/U_0$ decreases by approximately one order of magnitude.

The $U^{(1)}/U_0$ dependence on the Biot and Marangoni numbers directly involves the surfactant physicochemical hydrodynamics. When the Biot number is large, sorption kinetics is fast and the surfactant distribution in the Newtonian base state is nearly uniform. In this limit the fore-aft symmetry of the Hadamard-Rybczynski velocity is nearly preserved, and thus the first order Weissenberg correction necessary tends to zero. As the Biot number decreases, the surfactant distribution becomes more asymmetric as molecules collect at the trailing pole. With this large degree of assymetry in $\Gamma(\theta)$, the drag induced by the large Marangoni stress lowers the drop terminal velocity. The Marangoni dependence can be described as follows. For very large Marangoni number, the surfactant

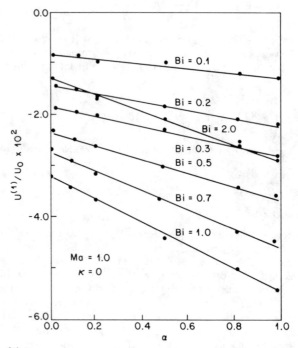

Figure 7: $U^{(1)}/U_0$ versus α for different Biot numbers and Marangoni number equal to one.

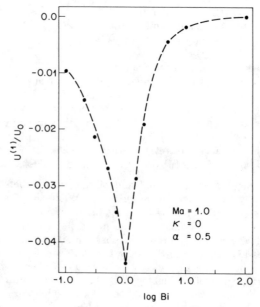

Figure 8: $U^{(1)}/U_0$ versus Biot number for Marangoni number equal to one.

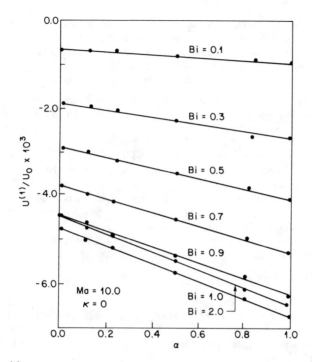

Figure 9: $U^{(1)}/U_0$ versus α for different Biot numbers and Marangoni number equal to ten.

distribution is uniform and the surface velocity is everywhere diminished. The result is a flow regime nearly identical to flow around a solid sphere. Because of the fore-aft symmetry this flow, $U^{(1)}/U_0$ approaches zero in agreement with the results of Leslie (1961). For smaller Marangoni number, the surface is more mobile and the influence of elongation drag slow drop translation.

The most interesting effect is that of viscoelasticity. For all values of α the correction is negative, despite the fact that for $\alpha=0$, no increase in the extensional viscosity is expected relative to that in Newtonian fluid. In this limit the non-Newtonian influence should be purely that of shear thinning. For vanishingly small α, lower viscosity would be expected to decrease the drag and thereby accelerate the drop. The fact that $U^{(1)}/U_0$ is negative, in this limit, can be explained by considering the secondary viscoelastic redistribution flow. The existence of a non-zero O(Wi) correction is a direct consequence of the break in flow symmetry around the equator. This asymmetry stems from two sources: 1) the zero order Marangoni stress, i.e. the Newtonian flow solution in the presence of surfactants, and 2) the distribution of surface adsorbed surfactant by the viscoelastic surface flow. The manner in which viscoelasticity enhances the Marangoni stress can be examined by considering the first order stream function in the absence of surfactants. The tangential velocity is of the form

$$v^{(1)}{}_\theta(r=1,Bi=0) = F(\epsilon,\alpha,\kappa)\sin\theta\cos\theta. \tag{31}$$

101

Note that this flow has a stagnation ring around the equator with the surface flow converging towards the poles. Thus in the presence of a viscoelastic continuous fluid, adsorbed surfactant is concentrated toward the trailing pole. This increased asymmetry in the surfactant distribution enhances the Marangoni stress and retards the drop translational velocity.

In this study we have examined effect of outer fluid viscoelasticity on the translational motion of a surfactant laden Newtonian droplet and we have found that the combined influence of fluid elasticity and secondary redistribution of surfactant play the major role in determining the hydrodynamic behavior in these systems. The effect of shear thinning (which has been previously examined by Kawase and Ulbrecht (1982) for a power-law fluid) is dominated by the increased extensional drag and interfacial gradient and thus the terminal velocity always decreases. Our findings suggest that understanding the drop hydrodynamics of non-Newtonian contaminated systems common to many industrial processes, requires rheological characterization which includes the elastic properties (i.e. dynamic measurements) in addition to shear viscosity measurements.

5. NONLINEAR SURFACE PRESSURE

The description of hydrodynamic retardation due to adsorbed surfactant requires an accurate representation of the Marangoni stresses. In most flow studies reported in the literature, it is assumed that the surface tension dependence on the local concentration can be represented by the gaseous expression:

$$\sigma_c - \sigma = RT\Gamma \tag{32}$$

where R and T denote the gas constant and the absolute temperature, and σ_c is the clean surface interfacial tension. This constitutive equation is restricted to low surface coverages. When the surfactant concentration per unit area is small, interactions between the adsorbed molecules are negligible, and the surface pressure dependence on Γ is analogous to the three-dimensional ideal gas law. As the surfactant density becomes large, the proximity of the molecules lead to repulsive forces within the monolayer. In this limit, the gaseous equation underpredicts the surface tension.

The use of a linear constitutive equation can lead to serious flaws for flow regimes in which the adsorbed surfactant density is high. If the viscous shear force exceeds the surface pressure force, contaminant molecules may be driven to trailing pole and highly compressed. In such a flow, σ will be larger than the gaseous value, and the extra tension must be described by a nonlinear equation. The effect of this extra surface pressure on drop hydrodynamics was recently examined by He, Dagan, and Maldarelli (Forthcoming) for stagnant cap flow. In this regime, adsorbed surfactant is driven to a stagnant region at the drop rear. The size of the immobilized cap region will depend on both the total amount of adsorbed surfactant, Γ_0, and on the extent of viscous compression. He et.al. predict larger cap size, ϕ, for a fixed Γ_0, than Sadhal and Johnson (1983) report. The assumption of gaseous surface tension behavior in the Sadhal and Johnson analysis underestimates the Marangoni forces and results in an underprediction of the drag coefficient. The deviation of the gaseous and nonlinear results depends on the relative strength of the viscous compressive forces as

102

characterized by the Marangoni number. When the surface pressure exceeds the viscous force, i.e. Ma $\gg$ 1, surfactant resists compression. He et.al. showed that flows characterized by large Marangoni numbers, exhibit substantial increases in ϕ for small increases in the amount of adsorbed surfactant. In this limit, the local concentration of surfactant remains low due the expanded cap size and the gaseous behavior adequately describes the Marangoni stresses. For small Marangoni numbers the behavior is quite different. In this case, the surface pressure forces are relatively weak and adsorbed surfactant is easily concentrated by the compressive viscous force. As a result of the higher local surface concentrations, strong repulsive forces exist on the interface which are inadequately modelled by the gaseous expression. In this analysis, we examine the effect of nonideal surface pressure behavior due to surfactant repulsive interactions, on the hydrodynamic motion of a surfactant covered drop for small but finite values of the Biot number. The translational terminal velocities which result from the nonlinear Marangoni stresses are determined and these are compared to the gaseous values. We first describe the nonlinear equation of state.

The Langmuir isotherm is used in conjunction with the Gibbs-Duhem equation to formulate an expression for the nonlinear surface pressure. In this isotherm, it is assumed that the surface concentration Γ_0 which is in equilibrium with the uniform bulk concentration C_∞ is given by

$$\Gamma_0 = \Gamma_\infty \frac{k}{1+k} \tag{33}$$

where Γ_∞ is the maximum packing density of the surfactant and k is the nondimensional bulk concentration, $k = \beta C_\infty / \alpha$. β and α are the adsorption and desorption rate constants with the net sorption given by

$$Q(\Gamma, C_s) = \beta C_s (\Gamma_\infty - \Gamma) - \alpha\Gamma \tag{34}$$

for a concentration of C_s, the bulk concentration of surfactant in the sublayer adjacent to the interface. The interfacial tension is related to the adsorbed surfactant concentration by the Gibbs-Duhem equation, $\Gamma = -RT\dfrac{\partial \sigma}{\partial \ln(C_s)}$, where the activity of surfactant in the bulk has been replaced by the bulk concentration. Integrating this equation with substitution of the Langmuir isotherm results in the Frumkin constitutive equation

$$\sigma_0 - \sigma_\infty = -RT\ln(1 - \Gamma/\Gamma_\infty)\Gamma_\infty . \tag{35}$$

When the surface coverage is low compared to the maximum packing density (i.e. $\Gamma \ll \Gamma_\infty$ or k small), the gaseous constitutive expression is recovered. The Frumkin surface pressure varies more strongly with Γ as the surface concentration increases (Figure 10). A serious drawback of the Frumkin equation is the predicted divergence of σ as the surface concentration approaches maximum packing density. In the high surface coverage limit, the Frumkin equation overcompensates for the repulsion between adsorbed molecules.

The shear stress difference across the spherical drop interface which results from the nonlinear equation of state depends on the both surface tension and on its gradient. For the Frumkin relation, the tangential boundary condition in dimensionless form is

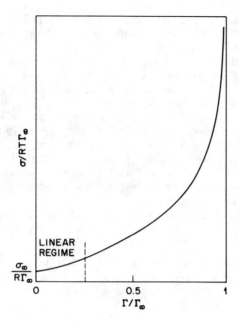

Figure 10: Frumkin surface pressure.

$$\tau_{r\theta} - \hat{\tau}_{r\theta} = \frac{\text{Ma}}{1-\Gamma} \frac{\partial \Gamma}{\partial \theta}. \tag{36}$$

The Marangoni number in this case is defined as $\text{RT}\Gamma_\infty /(\mu U_H)$, where the surfactant concentration has been nondimensionalized by the maximum packing concentration. Note that U_H is the Hadamard-Rybczynski terminal velocity. If we assume that diffusion of surfactant from the bulk to the interface occurs rapidly, then the bulk concentration of surfactant in the sublayer adjacent to the interface is constant, i.e. $C_s=1$, and the flux normal flux of contaminant molecules to the interface equals the net sorption flux. The surface mass conservation equation derived from the Langmuir adsorption isotherm is given by

$$\frac{1}{\sin\theta} \frac{\partial}{\partial \theta}(v_s(\theta)\Gamma\sin\theta) = \text{Pe}_s^{-1}(\frac{1}{\sin\theta} \frac{\partial}{\partial \theta}(\sin\theta\frac{\partial \Gamma}{\partial \theta})) - \text{Bi}\,(\text{k}(\Gamma - 1) + \Gamma) \tag{37}$$

The Biot number is defined as $\alpha a/U_H$.

The hydrodynamic solution is obtained by introducing the axisymmetric spherical stream functions into the governing drop and continuous phase momentum equations. The resulting fourth order homogeneous equations are solved subject to the following boundary conditions: 1) uniform free stream velocity, U, far away from the drop, 2) finite velocity within the drop, 3) continuity of the interfacial tangential velocity, 4) zero normal velocity at the drop surface, and 5) the tangential stress balance prescribed by equation (36). The Marangoni stress is a function of angular position only. Representation of the surface tension function $\sigma(\Gamma)$ as an infinite series of the form

104

$$\sum_{m=2}^{\infty} b_m \frac{C_m^{-1/2}(\cos\theta)}{\sin\theta} \tag{38}$$

permits solution of the inner and outer stream functions. Note that $C_m^{-1/2}(\cos\theta)$ is the Gegenbauer polynomial of degree $-1/2$ and order m. The interfacial tangential velocity as obtained from $\psi(r=1,\theta)$ is:

$$v_s = \frac{1}{1+\kappa} \sum_{n=2}^{\infty} \left[(1 + \frac{2Mab_1}{3(3\kappa+2)})\delta_{n2}Ma\ b_{n-1} \right] \frac{C_n^{-1/2}(\cos\theta)}{\sin\theta} \tag{39}$$

where κ is the ratio of drop to continuous phase viscosity. Substitution v_s and of $\Gamma = \sum_{n=0}^{\infty} a_n P_n(\cos\theta)$ into the surface mass balance equation (37) followed by orthogonalization i.e. the Galerkin weighted residual method (Finalyson, 1972), yield N coupled equations for b_n, where N is the number of coefficients retained in the truncated series. The Marangoni coefficients are determined numerically subject to the $\sigma(\Gamma)$ constraint

$$\sum_{n=2}^{N} n(n-1)a_{n-1} \frac{C_n^{-1/2}}{\sin\theta} \left[1 - \sum_{q=1}^{N} b_q P_q \right] = \sum_{m=2}^{N} m(m-1)b_{m-1} \frac{C_m^{-1/2}}{\sin\theta}. \tag{40}$$

The steady terminal velocity is obtained from a macroscopic balance of the viscous and the bouyant forces:

$$\frac{U}{U_H} = 1 + \frac{2Ma}{3(3\kappa+2)}b_1. \tag{41}$$

The translational drag of a surfactant covered gas bubble (i.e $\kappa = 0$) was determined for Bi ≤ 1 and several values of the Marangoni number. The influence of the total amount of adsorbed contaminant on the hydrodynamic retardation is examined through the terminal velocity dependence on the parameter k. The results are summarized in Figures 11 to 14.

The dimensionless steady translational velocity of a rising bubble ($\kappa=0$) is shown in Figure 11 for Ma equal to 1. U/U_H is plotted versus the Biot number for different Langmuir sorption ratios, that is k ranging from 10^{-2} to 10. With increasing k, the total amount of adsorbed contaminant increases until the maximum packing density is reached (Eq. 33). The presence of large amounts of surfactant hinders the mobility of the bubble interface and thus increasingly retards the terminal velocity as k rises. For k $= 10^{-2}$, the dimensionless translational velocity remains very close to the clean surface or the Hadamard-Rybczynski value of unity due to the restricted adsorption in the small k limit. A notable reduction from the clean surface velocity occurs for small k only in the limit as Bi $\rightarrow$ 0. In this stagnant cap flow regime, the surfactant accumulates at the drop rear and the correspondingly large interfacial tension gradient retards the drop motion. The stagnant cap terminal velocities, as determined by He et.al. (Forthcoming), are plotted on the y-axis; in the limit of vanishingly small Biot number our numerical results asymptotically approach these stagnant cap values (i.e the dotted curves). We have plotted the translational velocity which we calculated using the gaseous constitutive equation for k $= 10^{-2}$ and 10^{-1} (the dashed curves). These velocities were obtained by solving the governing equations in the limit as $\Gamma_0 \rightarrow k\Gamma_\infty$

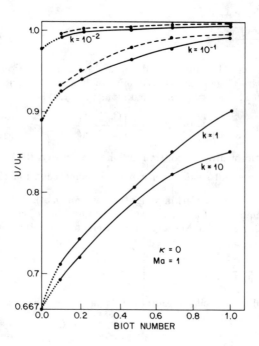

Figure 11: Steady terminal velocity as a function of the Biot number for different k values.

(dimensional) and with the quantity $(1-\Gamma)$ in the denominator of equation (36) replaced by unity. The linear and Frumkin constitutive equations predict nearly the same hydrodynamics. As k increases, however, use of the ideal gas formulation leads to serious underprediction of the terminal velocity as seen in Figure 12. In this plot, the bubble velocity is shown as a function of log k for different Biot numbers. For a fixed Bi, U/U_H decreases with increasing k. The linear and nonlinear hydrodynamic behavior is compared for Bi = 0.5. Good agreement is predicted (with the gaseous constitutive equation formulation slightly overpredicting the steady state bubble velocity) for k less than 0.2. For higher k values, U/U_H drops off rapidly as a function of k, and the linear results severly underpredict the translational motion.

In Figures 13 and 14, the Marangoni number dependence of U/U_H is examined for Bi equal to one. The terminal velocities are plotted as a function of log Ma for several k values in Figure 13. With increasing Ma, the Marangoni tension exerted on the bubble increases, and thereby leads to translational velocity reductions. For fixed Marangoni number, U/U_H decreases with increases k, due to the greater interfacial mobility hindrance associated with increased Γ_0. For the two smaller k values plotted here, the hydrodynamic predictions obtained from the ideal gas formulation are in good agreement with the Frumkin results, in that they only slightly overpredict the terminal velocity. For both k equal to 1 and 10, the situation is reversed. This can be seen in Figure 14 in which we plot U/U_H versus log k for Bi=1. The reduction in the translational velocity with increasing k, is shown for Ma = 0.1 ,1 and 10. A comparison of the gaseous

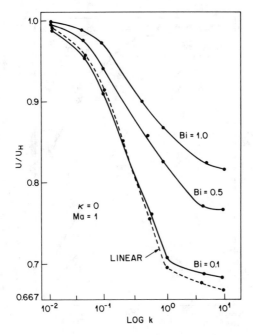

Figure 12 : Steady terminal velocity as a function of log k for different Biot numbers.

and Frumkin predictions for the drop hydrodynamics are shown for Ma=1. When k is approximately 10^{-1} or less, the gaseous results agree well with nonlinear terminal velocity values. For larger k, the hydrodynamic predictions diverge, with the linear formulation seriously underpredicting the droplet velocity.

Drop translational motion in the presence of surfactants exhibits a strong dependence on the adsorption kinetics and on the interfacial tension equation of state assumed in the problem formulation. Our results show that while the assumption of gaseous surface pressure behavior is adequate when the concentration of adsorbed surfactant is low, serious flaws in the hydrodynamic predictions result if Γ_0 exceeds 10% of the maximum packing density. This behavior can be explained as follows. In formulating the contaminated Stokes flow problem with the assumption of Langmuir adsorption kinetics and the Frumkin constitutive equation, the equilibrium concentration of adsorbed surfactant is $\Gamma_0=k/k+1$. The Langmuir kinetics thus predict that Γ_0 approaches the maximum packing density, for k increasing, but cannot exceed it. The kinetics of the linear formulation, however, does not account for the physical limit to adsorption derives from the repulsive interactions which developed during close molecular packing at the surface. As a result, the ideal gas formulation leads to $\Gamma_0=k$, and thus for large k, i.e. C_∞ large, the Γ_0 is erroneously predicted to exceed Γ_∞. Owing to the fact that the linear adsorption kinetics overpredicts the amount of surfactant adsorption which is physically realizable for a fixed C_∞, the linear terminal velocities overpredict

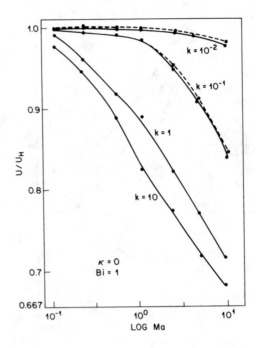

Figure 13: Steady terminal velocity as a function of the logarithm of the Marangoni number for different k values.

the extent of surface immobilization and thereby underpredicts U/U_H. For k small, the slight discrepancy between the gaseous and Frumkin results can be attributed to the difference in the Marangoni stress. Although Γ_0 values are virtually identical, the presence of $(1 - \Gamma)$ in the denominator of the Frumkin Marangoni tension, lead to larger Marangoni tension (i.e. since Γ is always less than or equal to one) and therefore to a higher viscous drag.

Accurate simulation of the Marangoni influence of surfactant adsorption on drop or bubble motion requires a realistic physical description of the adsorption kinetics and of the interfacial pressure dependence on the Γ. Despite the widespread use of the ideal gas formulation for $\sigma(\Gamma)$, our numerical results reveal that the inability of this constitutive model to account for finite adsorption due to intermolecular repulsion can lead to critical flaw in the predicted hydrodynamics.

6. CONCLUSIONS

We have reviewed some recent results for single drop mass transport and hydrodynamics for low Reynolds number flow in the presence of surfactants. The large body of literature concerning the influence of surfactants on dropwise mass transport and the increasing number of studies concerning drop hydrodynamics in non-Newtonian fluids indicate the importance of dropwise

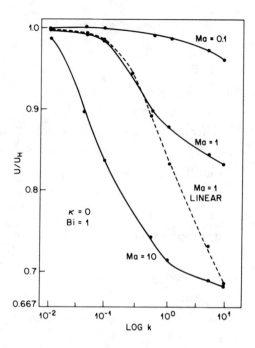

Figure 14: Steady terminal velocity as a function of log k for different Marangoni numbers.

transport for many industrial multiphase processes. Despite the existence of a large number of papers in the area of surfactant Marangoni flows, academic interest persists because of our lack of understanding of many of the subtle and practical aspects of this class of problems. In this paper, we examined the effect of surfactant steric blockage on interfacial transport and the retarding influence of fluid elasticity on drop motion under the assumption that the interfacial tension is proportional to the surfactant concentration. The relaxation of the ideal gas assumption, and the solution of the hydrodynamic equations using the Frumkin equation of state, show that repulsive interactions between adsorbed molecules strongly influence the drop motion. In light of this result, reexamination of interfacial mass transport, drop hydrodynamics in a viscoelastic medium and other transport process which take place in the presence of contaminants is warranted, via simulations which incorporate molecular repulsion.

NOMENCLATURE

a	drop radius
a_n	surfactant concentration coefficients, $0 \leq n \leq \infty$
$a_n^{(k)}$	k^{th} order surfactant concentration coefficients, $0 \leq n \leq \infty$
A_n, B_n, C_n, D_n	k^{th} order outer fluid stream function coefficients, $0 \leq n \leq \infty$
b_n	surface tension gradient coefficients, $0 \leq n \leq \infty$
Bi	Biot number, ha/U
$c\,(r, \theta)$	solute concentration
$C\,(r, \theta)$	dimensionless solute concentration
c_i	solute concentration in outside sublayer near drop
$C_n^{-1/2}(\cos\theta)$	n^{th} order Gegenbauer polynomial of degree -1/2
c_o	disperse phase solute concentration
c_∞	solute concentration far away from drop
C_∞	surfactant concentration far away from drop
Ca	Capillary number, $\mu U/\sigma$
D	continuous phase solute diffusion coefficient
D_s	surface diffusion coefficient
$\dfrac{D}{Dt}$	corotational objective time derivative
$\dot{E}$	elongation rate, (sec^{-1})
h	sorption rate (sec^{-1})
k	nondimensional surfactant bulk concentration, $\beta C_\infty / \alpha$
Ma	Marangoni number, $RT\Gamma_0/\mu U$
$P_k(\cos\theta)$	k^{th} order Legendre polynomial
Pe	continuous phase Peclet number, $2U_H a/D$
Pe_s	surface Peclet number, aU_H/D_s
r	radial coordinate
Re	Reynolds number, $\rho a U_H/\mu$
Sh	Sherwood number
$v_r\,(r, \theta)$	dimensionless continuous phase radial velocity
v_s	interfacial tangential velocity
$U_s(\phi)$	stagnant cap drop terminal velocity
U_H	solid sphere terminal velocity

$v_\theta\,(r,\,\theta)$	dimensionless continuous phase tangential velocity
Wi	Weissenberg number, $\lambda_1 U_H / a$
$Z(r,\theta)$	non-Newtonian part of the stream function equation

Greek symbols

α	variable objective time derivative parameter
α,β	Langmuir sorption rate constants
ϵ	ratio of retardation to relaxation times, λ_2 / λ_1
$\dot{\gamma},\hat{\dot{\gamma}}$	continuous and disperse phase rate of strain tensors
$\Gamma\,(\theta)$	interfacial surfactant concentration
Γ_0	equilibrium surfactant concentration
Γ_∞	maximum packing surfactant concentration
κ	ratio of drop to continuous phase viscosity
λ_1,λ_2	relaxation, retardation times
$\mu,\hat{\mu}$	continuous and disperse phase viscosity
η_0	zero shear rate viscosity
$\eta,\overline{\eta}$	shear, elongational viscosity
$\Omega(\tau,\dot{\gamma},\kappa,\alpha,\epsilon,\mathrm{Wi})$	non-Newtonian stress tensor
ϕ	stagnant cap angle
σ	interfacial tension
σ_c	clean surface interfacial tension
$\tau,\hat{\tau}$	continuous and disperse phase stress tensor
θ	tangential coordinate
$\psi,\hat{\psi}$	continuous and disperse phase stream functions

REFERENCES

Archer, R. J. and La Mer, V., The rate of evaporation through fatty acid monolayers, *J. Phys. Chem.* **59**, 200-208 (1955).

Beitel, A. and Heideger, W. J., Surfactant effects on mass transfer from drops subject to interfacial instability, *Chem. Eng. Sci.* **26**, 711-717 (1971).

Burnett, J. C. and Himmelblau, D. M., The effect of surface active agents on interphase mass transfer, *AIChE J.* **16**, 185-193 (1970).

111

Caswell, B. and Schwarz, W. H. The creeping motion of a non-Newtonian fluid past a sphere, *J. Fluid Mech* **13**, 417-426 (1962).

Davis, R. E. and Acrivos, A., The influence of surfactants on creeping motion of bubbles, *Chem. Eng. Sci.* **21**, 681-685 (1966).

Elzinga, E. R. and Banchero, J. T., Some observations on the mechanics of drops in liquid-liquid systems, *AIChE J.* **7**, 394-399 (1961).

Finlayson, B, A., The Method of Weighted Residuals and Variational Principles, pp. 88-91, Academic Press, New York, 1972.

Frumkin, A. N. and Levich, V. G., Effect of surface-active substances on movement at the boundaries of liquid phases, *Zhur. Fiz. Khim.* **21**, 1183-1204 (1947).

Garner, F. H. and Hale, A. R., The effect of surface active agents in liquid extraction processes, *Chem. Eng. Sci.* **2**, 157-163 (1953).

Garner, F. H. and Skelland, A. H. P., Some factors affecting droplet behavior in liquid-liquid systems, *Chem. Eng. Sci.* **4**, 149-158 (1955).

Giesekus, H., Die simultane Translations- und Rotationsbewegung einer Kugel in einer elastoviskosen Fluessigkeit, *Rheol. Acta* **3**, 59-70 (1963)

Griffith, R. M., The effects of surfactants on the terminal velocity of drops and bubbles, *Chem. Eng. Ser.* **17**, 1057-1070 (1962).

Happel, J. and Brenner, H., Low Reynolds Hydrodynamics, pp. 133-137, Prentice-Hall, Englewood Cliffs, New Jersey (1965).

Harper, J. F., On bubbles with small immobile adsorbed films rising in liquids at low Reynolds number, *J. Fluid Mech.* **58**, 539-545 (1973).

Harper, J. F., Surface activity and bubble motion, *Appl. Sci. Res.* **38**, 343-352 (1982).

He, Z., Dagan, Z. and Maldarelli, C. M., The influence of surfactant on the motion of a fluid sphere in a tube. Part II. The stagnant cap regime, *J. Fluid Mech.* (Forthcoming).

Holbrook, J. A. and Levan, M. D., Retardation of droplet motion by surfactant. Part 2. Numerical solutions for exterior diffusion, surface diffusion, and adsorption kinetics, *Chem. Eng. Comm.* **20**, 273-290 (1983).

Horton, T. J., Fritsch, T. R., and Kintner, R. C., Experimental determination of circulation velocities inside drops, *Can. J. Chem. Eng.* **43**, 143 (1965).

Huang, W. S. and Kintner, R. C., Effect of surfactants on mass transfer inside drops, *AIChE J.* **15**, 735-744 (1969).

Kawase, Y., and Ulbrecht J. J., The effect of surfactant on terminal velocity of and mass transfer from a fluid sphere in a non-Newtonian fluid, *Can. J. Chem. Eng.* **60**, 87-93 (1982).

Langmuir, I. and Schaefer, V. J., Rates of evaporation of water through compressed monolayers on water, *J. Franklin Inst.* **235**, 119-162 (1943).

Leslie, F. M., The slow flow of a visco-elastic liquid past a sphere, *Quart. J. Mech. and App. Math.* **14** 36-49 (1961).

Levan, M. D. and Newman, J., Effect of surfactant on the terminal and interfacial velocities of a bubble or drop, *AIChE J.* **22** 695-701 (1976).

Levich, V. G., Physiochemical Hydrodynamics, p. 390. Prentice-Hall, New York (1962).

Lindland, K. P. and Terjesen, S. G., The effect of a surface-active agent on mass transfer in falling drop extraction, *Chem. Eng. Sci.* **5**, 1-12 (1956).

Lochiel, A. C., The influence of surfactants on mass transfer around spheres, *Can. J. Chem. Eng.* **43**, 40-44 (1965).

Mekasut, L., Molinier, J. and Angelino, H., Effects of surfactants on mass transfer outside drops, *Chem. Eng. Sci.* **33**, 821-829 (1978).

Mudge, L. K. and Heideger, W. J., The effect of surface active agents on liquid-liquid mass transfer rates, *AIChE J.* **16**, 602-608 (1970).

Plevan, R. E. and Quinn, J. A., The effect of monomolecular films on the rate of gas adsorption into a quiescent liquid, *AIChE J.* **12**, 894-902 (1966).

Quintana, G. C., Ph.D. Thesis, Columbia University, 1986.

Quintana, G. C., Cheh, H. Y. and Maldarelli, C. M., The translation of a Newtonian droplet in a 4-constant Oldroyd fluid, *J. Non-Newt. Fluid Mech.* **22** 253-270 (1987).

Rideal, E. K., On the influence of thin surface films on the evaporation of water, *J. Chem. Phys.* **29**, 1585-1588 (1925).

Sadhal, S. S. and Johnson, R. E., Stokes flow past bubbles and drops partially coated with thin films. Part I. Stagnant cap of surfactant film - exact solution, *J. Fluid Mech.* **126**, 237-250 (1983).

Savic, P., Circulation and distortion of liquid drops falling through a viscous medium. Nat. Res. Counc. Can. Div. Mech. Eng. Rep. MT-22 (1953).

Tiefenbruck, G. F., Ph.D. Thesis, California Inst. of Technology, 1980.

Chapter 5

Rate of Gas Absorption by Liquids and "Surface Resistance"

P. TIKUISIS
Defence and Civil Institute of Environmental Medicine
1133 Sheppard Avenue West
Downsview, Ontario M3M 3B9, Canada

C. A. WARD
Thermodynamics and Kinetics Laboratory
Department of Mechanical Engineering
University of Toronto
5 King's College Road, Toronto M5S 1A4, Canada

ABSTRACT

The common procedure used to determine the diffusion coefficient (D) of a gas in a liquid is to expose the liquid to the gas for a period of time, measure the amount of gas absorbed and then determine D from a fitting procedure. It is normally assumed that upon exposure, the interphase is instantaneously equilibrated with the gas phase, but this assumption has led to inconsistencies in the values of D that are inferred from different experimental techniques. In the case of oxygen dissolving in water, this procedure has led to very small values of D being inferred when the exposure time is brief, as compared to the values of D obtained when the experimental technique involves longer exposure times. By contrast, when carbon dioxide is dissolving in water, experimental techniques that involve different periods of exposure nonetheless lead to consistent values of the diffusion coefficient. The discrepancy in the value of D has been suggested to be caused by "surface resistance". If Statistical Rate Theory is adopted to predict the rate of gas absorption at the liquid-gas interface, the equilibrium assumption can be replaced by an expression for the rate of gas absorption. This rate expression only involves parameters that may be evaluated by independent experimental techniques and when it is applied to examine the experimental data for both of these gases, it is found that a consistent value of the diffusion coefficient is obtained for each of the gases. The results suggest that Statistical Rate Theory gives an explanation for the phenomena that has been associated with "surface resistance".

INTRODUCTION

When an initially pure liquid phase is exposed to a gas component at a phase boundary (see Fig. 1), a series of processes is initiated that ultimately leads to the gas component being absorbed into the bulk of the liquid. These processes may be idealized as consisting of three sequential steps. The first is the transfer of gas molecules from the gas phase to the interphase, followed by the transfer of some of them from the interphase to the bulk liquid and then their diffusion further into the bulk [1–4]. The net rate of these first two steps acts as the boundary condition for the diffusion equation

$$(\partial c / \partial t) = D \nabla^2 c \tag{1}$$

where c is the gas concentration in the bulk liquid.

Historically, the rate of gas absorption by a liquid phase has not been viewed as a three-sequential rate processes [5]. The usual assumption applied at the boundary of the liquid phase is that this interphase goes immediately to equilibrium when it is exposed to the gas [6]; thus

$$c(0,t) = \alpha \qquad\qquad t > 0 \qquad\qquad (2)$$

where α is the solubility.

If the Henry relation is used to describe the solubility of the gas in the liquid solvent, then α is given by

$$\alpha = c_1 P_g / K_H \qquad\qquad (3)$$

where c_1 is the concentration of the liquid solvent. P_g is the partial pressure of the gas component in the gas-vapour mixture contacting the liquid, and K_H only depends on temperature.

If the diffusion into the liquid phase may be treated as one-dimensional, one then finds from Eqs. (1) and (2)

$$c(z,t) = c_i + (\alpha - c_i)(erfc\ (z/2\sqrt{Dt}\ ) \qquad\qquad (4)$$

where c_i is the initial concentration of the gas component in the liquid and z is the spatial variable. The flux of the gas component across the phase boundary, J_{GL}, is given by

$$J_{GL} = -D\ (\partial c / \partial z)_o \qquad\qquad (5)$$

and one finds from Eq.(4) that

$$J_{GL} = \left(\alpha - c_i\right)\sqrt{D/\pi t} \qquad\qquad (6)$$

One may then integrate Eq. (6) to determine the amount absorbed in the period from zero to a time t.

$$Q(t) = 2(\alpha - c_i)\left(\sqrt{Dt/\pi}\ \right) \qquad\qquad (7)$$

To determine the effect of factors other than diffusion on the rate of gas absorption, one compares the measured amount of absorption with that predicted from Eq. (7). Note that in order for Eq. (7) to be valid, the rate of molecular transport from the gas phase to the liquid phase must be initially infinite (see Eq. (6)). It is recognized that this singularity is unreasonable; however the argument revolves around whether it places an important restriction of Eq. (7) [6, 7].

If the absorption measured is less than that predicted from Eq. (7), then it is usually attributed to a "surface resistance". However, at this point, the argument becomes a little complex because whether it is concluded that there is an important surface resistance acting or not depends on the value of the diffusion coefficient used in the calculation and the value of this parameter is usually inferred by assuming the surface resistance to be negligible.

That this procedure leads to inconsistencies can be seen from the values of the diffusion coefficient that are inferred from different experimental techniques. For example, for O_2 dissolving into water, the diffusion coefficient inferred by Chiang and Toor [8] under the assumption of local equilibrium from data obtained with a liquid-jet technique was $1.6 \times 10^{-5}\ cm^2/s$ at 22.2°C. The exposure time in their experiments ranged from 0.8 to

11.8 *ms*. This value of the diffusion coefficient is so much lower than other values reported in the literature that Chiang and Toor concluded reluctantly that, in their experimental circumstance, significant surface resistance effects must have been involved.

More recent values of the diffusion coefficient obtained from experimental techniques that exposed water to O_2 under approximately steady state conditions have led to even larger values of the diffusion coefficient than Chiang and Toor considered. For example, Ng and Walkley [9] used a technique in which the O_2 flow rate required to maintain a bubble at a constant size was measured. The bubble was maintained at a particular size for a period of approximately 3,000 *s,* and they inferred a value of D that was 65% higher than the value that Chiang and Toor obtained. Thus the reluctant conclusion of Chiang and Toor of surface resistance would have been *a fortiori* in view of this larger value of the diffusion coefficient.

One of the reasons that Chiang and Toor and others [6–7] have been reluctant to conclude that there is an important surface resistance for the case of O_2 absorption in water is because of the experience with the CO_2 water system. Initial studies by Higbie [10] suggested that there could be an important surface resistance for this system and Danckwerts and Kennedy [11] initially found a similar result; however, the latter two investigators later decided that their results could be accounted for by a smaller value of the diffusion coefficient [12]. Then Raimondi and Toor [13] conducted a study with liquid jets in which the exposure time was on the order of a millisecond and using values of the diffusion coefficient that had been determined from other experimental techniques with much longer exposure times, concluded that there was no significant surface resistance in the case of the CO_2-water system.

If one proposes to explain the source of a surface resistance for the case of O_2-water, it would appear from the experimental evidence that the theory must also explain the reason there is no surface resistance for the CO_2-water system. A theory has been previously proposed by Ward [3] to explain the reason for the surface resistance in the case of O_2-water. The theoretical approach, called Statistical Rate Theory, is based on statistical mechanics and uses the concept of transition probabilities to predict the rate of molecular transport across an interface [1,14]. When this approach is applied to obtain the expression for the rate of O_2 absorption at a gas-water interphase and this rate expression is coupled with the diffusion equation to interpret the steady state gas flow data of Ng and Walkley [9] and the liquid-jet data of Chiang and Toor [8], it is found that no adjustment of the diffusion coefficient is necessary.

In the present study, a review of the development of Statistical Rate Theory is given followed by its use to determine the molecular absorption cross-sections of O_2 and CO_2 from the liquid-jet data of Chiang and Toor [8] and Raimondi and Toor [13], respectively. In these cases, the value of the diffusion coefficient is adopted from the steady state values reported by Ng and Walkley [9]. It is found that the value of the molecular absorption cross-section of O_2 that is inferred from Statistical Rate Theory is in close agreement with tabulated values of the molecular radius of O_2. The value inferred for CO_2 is consistent with the known reaction of CO_2 and water [15]. Finally, it is shown that a parameter can be defined that describes when the expression for the rate of gas absorption must be taken into account and when local equilibrium may be assumed to exist at the liquid-gas interphase. This non-dimension parameter (v) depends on the properties of the liquid-gas combination and the exposure time of the particular experimental technique. In the case of the measured absorption of O_2 into H_2O, it is found that for the exposure times of the liquid-jet, the local equilibrium assumption can not be made without error. Whereas for C_2O absorbing into H_2O, one finds that the local equilibrium assumption is valid for the liquid-jet data that has been reported in the past. However, if the exposure time were reduced by approximately an order of magnitude, the rate of CO_2 absorption would have to be taken into account as well.

REVIEW OF STATISTICAL RATE THEORY

Consider an isolated system that consists of two components that are distributed between three phases, a liquid, a gas and the interphase, denoted by the superscripts L, G, and I, respectively (see Fig. 1). Let N_i^j be the number of molecules of component i that are in phase j. At the instant t_i, the distribution of molecules between the phases will be denoted as λ_i where

$$\lambda_i : N_{1i}^G, N_{1i}^I, N_{1i}^L, N_{2i}^G, N_{2i}^I, N_{2i}^L \tag{8}$$

Following the Statistical Rate Theory approach, we propose to predict the probability of a change in the molecular configuration in which one gas molecule (component 2) is transferred from the gas phase to the interphase so that in the time interval δt, the molecular configuration becomes λ_j where

$$\lambda_j : N_{1i}^G, N_{1i}^I, N_{1i}^L, N_{2i}^G - 1, N_{2i}^I + 1, N_{2i}^L \tag{9}$$

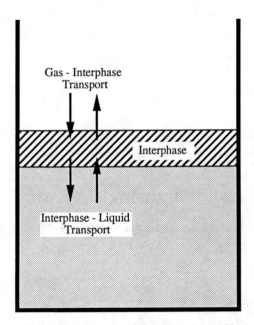

FIGURE 1. Schematic of gas absorption problem showing three distinct phases.

To calculate the probability of this transition occurring at any instant, we may employ a standard quantum mechanical approximation, first order perturbation theory [16]. After introducing the hypothesis that the rate of transition between quantum mechanical states is equal to the average rate for all states within the energy uncertainty, one finds [1, 14] as the probability of a transition at any instant

$$\tau(\lambda_i, \lambda_j) = K_{GI} \, \Omega(\lambda_j) / \Omega(\lambda_i) \tag{10}$$

where $\Omega(\lambda)$ is the number of quantum mechanical states corresponding to molecular configuration λ and K_{GI} is the equilibrium exchange rate at the interface between the gas and the interphase. Note that the value of the latter rate is unknown, but its value is fixed by the constraints of the system just as the final equilibrium configuration is fixed by the constraints. The equilibrium configuration can, for example, be predicted from thermodynamics. We shall consider the expression for K_{GI} in detail subsequently. If $\tau(\lambda_i, \lambda_j)$ is the probability of the system making a transition in which one molecule is transferred from the gas phase to the interphase, then the number of transitions, $\delta\eta$, in the time interval δt is approximately

$$\delta\eta = \tau(\lambda_i, \lambda_j) \, \delta t \qquad (11)$$

and the rate of transition from λ_i to λ_j is

$$J(\lambda_i, \lambda_j) = K_{GI} \, \Omega(\lambda_j) \, / \, \Omega(\lambda_i) \qquad (12)$$

At the same instant that there is a probability for a transition from λ_i to λ_j, there is a probability for a transition in which a gas molecule is transferred from the interphase to the gas phase, i.e. from λ_i to λ_h where

$$\lambda_h: \; N_{1i}^G, N_{1i}^I, N_{1i}^L, N_{2i}^G + 1, N_{2i}^I - 1, N_{2i}^L \qquad (13)$$

For this rate process, one finds

$$J(\lambda_i, \lambda_h) = K_{GI} \, \Omega(\lambda_h) \, / \, \Omega(\lambda_i) \qquad (14)$$

The net rate at which gas molecules are absorbed in the interphase, J_{GI}, is then the difference between $J(\lambda_i, \lambda_j)$ and $J(\lambda_i, \lambda_h)$ and after introducing the Boltzmann definition of entropy, the net rate may be written

$$J_{GI} = K_{GI} \left[\exp\left((S(\lambda_h) - S(\lambda_i)) / k \right) - \exp\left((S(\lambda_j) - S(\lambda_i)) / k \right) \right] \qquad (15)$$

We may use thermodynamics to simplify the expression for the change in entropy. For any molecular distribution λ_k, we may write

$$S(\lambda_k) = S^G(\lambda_k) + S^I(\lambda_k) + S^L(\lambda_k) \qquad (16)$$

and for the bulk phases, the Euler relation gives

$$S = (1/T) \, U + (P/T) \, V - (N_1 \mu_1 / T + N_2 \mu_2 / T) \qquad (17)$$

and for the interphase

$$S^I = (1/T) \, U^I - (\gamma^I / T) \, A^I - (N_1^I \mu_1^I / T + N_2^I \mu_2^I / T) \qquad (18)$$

When these relations are substituted into Eq. (16), and it is assumed that the transfer of one molecule between phases does not change the intensive properties, one finds for the entropy changes

$$S(\lambda_h) - S(\lambda_i) = (\mu_2^I - \mu_2^G)/T \tag{19}$$

and

$$S(\lambda_j) - S(\lambda_i) = (\mu_2^G - \mu_2^I)/T \tag{20}$$

Then the expression for the net rate of molecular transfer from the gas phase to the interphase is given by

$$J_{GI} = K_{GI} \sinh\left[(\mu_2^G - \mu_2^I)/kT\right] \tag{21}$$

When the same procedure outlined above is used to obtain the expressions for the net rate of molecular transfer from the interphase to the liquid phase, one finds

$$J_{IL} = K_{IL} \sinh\left[(\mu_2^I - \mu_2^G)/kT\right] \tag{22}$$

where K_{IL} is the equilibrium rate of molecular exchange across the boundary between the interphase and the bulk liquid phase.

If conservation of mass at the interphase is now imposed, we may write that

$$(1/A^I)(dN_2^G/dt) = J_{GI} \tag{23}$$

$$(1/A^I)(dN_2^I/dt) = J_{GI} - J_{IL} \tag{24}$$

$$-D(\partial c/\partial z)_o = J_{IL} \tag{25}$$

In principal, these equations could now be solved simultaneously with the diffusion equation if the expression for μ_2^G, μ_2^I, μ_2^L, K_{GI}, and K_{IL} were known in terms of N_2^I, N_2^G, and $c(0,t)$. However, the expression for μ_2^I is not known and, further, our primary interest lies in predicting the amount of component 2 absorbed by the liquid phase.

To simplify the analysis, we shall assume that the number of gas molecules that are in the interphase is at all times negligible. Thus the gas molecules pass through the interphase but their accumulation there is negligible. Then

$$J_{GI} = J_{IL} \tag{26}$$

and the slower of the two rates will be rate-limiting. First, we consider the possibility that J_{GL} is small compared to J_{IL}, then the I-L interface must be near equilibrium; i.e.

$$\mu_2^I = \mu_2^L \tag{27}$$

and the expression for the limiting rate J_L may be simplified to give

$$J_{GI} = J_L = K_{GI} \sinh\left[(\mu_2^G - \mu_2^L)/kT\right] \tag{28}$$

119

Now consider the other possibility. Suppose J_{IL} is small compared to J_{GI}, then in this case, the G - I interface would be near equilibrium

$$\mu_2^G = \mu_2^I \tag{29}$$

and the limiting rate in this case would be J_{IL}, expressed as

$$J_{IL} = J_L = K_{IL} \sinh\left[(\mu_2^G - \mu_2^L) / kT\right] \tag{30}$$

If one compares the two possible expressions for the limiting rate, it may be noted that these expressions differ only by the factor multiplying the hyperbolic sine term, i.e. equilibrium exchange rate for the rate-limiting step. Of the two possible rate-limiting steps in the sequence, the one that actually limits the rate then will be the one that has the smallest equilibrium exchange rate [1].

To determine the expression for the equilibrium exchange rates, we shall adopt a procedure similar to that described in Refs. [2] and [3] and write them in the form

$$K_{ij} = v_{ij} (2) A_{ij} (2) \tag{31}$$

where K_{ij} is the equilibrium exchange rate at the i - j interface, i.e. G-I or I-L interface; v_{ij} is the molecular collision frequency of component 2 molecules coming from phase i with the i-j interface and A_{ij} (2) is the fraction of the i-j interface that is available for these molecules to be transferred into phase j.

For the component 2 molecules coming from the interphase and entering the liquid phase, there is considerable impedance because many of the absorption sites are occupied under equilibrium conditions. To estimate the total number of absorption sites that could exist in the liquid phase, we first estimate the maximum concentration of component 2 molecules, c_s, that could be present there if the liquid phase were at a pressure P and temperature T. We take c_s to be the saturation concentration and assume that it can be calculated from the Henry relation. For a liquid at a total liquid pressure of P, we have

$$c_s = c_1 P / K_H \tag{32}$$

Under equilibrium conditions, the concentration of molecules absorbed, α, would be

$$\alpha = c_1 (P - P_\infty) / K_H \tag{33}$$

since the liquid phase would be in contact with a gas-vapour phase and the partial pressure of the gas would be $P - P_\infty$, where P_∞ is the vapour pressure. The number of absorption sites per unit area that are within a molecular radius then could be estimated as the difference in c_s and c_e multiplied by the molecular absorption cross-sectional radius (σ) and the area of the sites would be approximately this number multiplied by the absorption cross-section area of the molecules; thus A_{IL} (2) would be given by

$$A_{IL} (2) = (c_s - \alpha) \pi \sigma_2^3 \tag{34}$$

In Ref. [3], it is argued that v_{IL} (2) is approximately equal to v_{GI} (2). Under equilibrium conditions, the gas phase will be in a Maxwellian distribution and the latter collision frequency will be

$$v_{GI}(2) = \frac{(P - P_\infty)}{\sqrt{2\pi m_2 kT}}$$
(35)

where m_2 is the mass of the component 2 molecule. Thus with these approximations, the equilibrium exchange rate at the $I - L$ interface may be expressed

$$K_{IL} = \left[(c_1 P_\infty)(P - P_\infty) \pi \sigma_2^3 / K_H \sqrt{2\pi m_2 kT} \right]$$
(36)

At the $G - I$ interface, there is little impedance to the transport of molecules from the gas phase to the interphase and $A_{GI}(2)$ is thus approximately unity [2]; thus

$$K_{IL} / K_{GI} = A_{IL}(2)$$
(37)

and since $A_{IL}(2)$ is small compared to unity, the rate-limiting process is the transfer of molecules from the interphase to the liquid.

Since we have limited our attention to weak liquid-gas solutions, we may approximate the expression for the chemical potential of the dissolved gas in the liquid phase as [17]:

$$\mu_2^L (T, P, c) = \mu_2^o (T.P) + kT \ln (c(0,t)/c_s)$$
(38)

where $\mu_2^o (T, P)$ is the chemical potential of the pure gas component. We shall approximate the gas phase as being ideal, then the well known expression for the chemical potential of component 2 in the gas phase is

$$\mu_2^G = \mu_2^o (T,P) + kT \ln x_2^G$$
(39)

where x_2^G is the mole fraction of component 2 in the gas phase. After subtracting μ_2^L from μ_2^G and substituting into Eq. (30), one finds as the expression for the limiting rate

$$J_L = K_{GL} \left[\left[\alpha / c(0,t) \right] - \left[c(0,t)/\alpha \right] \right]$$
(40)

where, as may be seen from Eqs. (32) and (33), α may be written

$$\alpha = x_2^G c_s$$
(41)

The expression for the rate of gas absorption at any instant at the boundary of the liquid phase is now complete.

AMOUNT OF GAS ABSORPTION BY A LIQUID PHASE

Suppose a liquid phase that is initially pure is suddenly exposed to a gas phase that consists of the vapour of the liquid and of a gas component that forms a weak solution with the liquid component. Also suppose that the diffusion problem may be described as one dimensional (see Fig. 1). We shall assume that the absorption of the gas occurs isothermally, since there will be little vaporization or condensation. The boundary value problem to be solved may now be stated:

$$(\partial c \,/\, \partial t) \;=\; D \,(\partial^2 c \,/\partial z^2) \tag{42}$$

$$-D \,(\partial c \,/\, \partial z)_o \;=\; K_{IL} \left[\left(\alpha \,/\, c\,(0,t) \right) - c\,(0,t) \,/\, \alpha \right] \tag{43}$$

where $c\,(o, t)$ is the concentration at the boundary of the liquid phase, and

$$c\,(z,0) = 0 \qquad\qquad z > 0 \tag{44}$$

The amount absorbed at the time t_e can be written in terms of J_L, i.e.

$$Q\,(t_e) \;=\; \int_o^{t_e} J_L \; dt \tag{45}$$

and after non-dimensionalizing the system of equations by introducing the non-dimensional concentration

$$u \;=\; (c \,/\, \alpha) \tag{46}$$

the non-dimensional time

$$v \;=\; (t\, K_{IL}^2 /\, \alpha^2 D) \tag{47}$$

and non-dimensional spatial variable y as

$$y \;=\; (z\, K_{IL} /\, \alpha D) \tag{48}$$

one finds

$$(\partial u \,/\, \partial v) \;=\; (\partial^2 u \,/\partial y^2) \tag{49}$$

$$(\partial u \,/\, \partial y)_o \;=\; \left[u\,(0,v) - 1\,/\,u\,(0,v) \right] \tag{50}$$

and

$$Q\,(v) = (\alpha^2 \, D \,/\, K_{IL}) \int_o^v \left[\left(1\,/\,u\,(0,v') \right) - u\,(0,v') \right] dv' \tag{51}$$

Note that the integral equation for $Q(v)$ has a singularity at the initial time because at this time $u(o, v)$ vanishes. This singularity is integrable, but it means that the concentration changes rapidly during the initial period. However, at longer times when u approaches unity, i.e. when the interphase approaches equilibrium, $Q(v)$ changes slowly with time. Thus we shall divide the problem into two time regimes. To obtain the short-time solution, we shall follow a procedure introduced by Peters [18].

First, we take the Laplace transform of Eq. (49) and obtain

$$\partial^2 \zeta \,/\partial y^2 - s \;=\; -\zeta \; ; \tag{52}$$

where the Laplace transform of u is

$$\pounds \, [u] \ = \ \zeta \tag{53}$$

and

$$\zeta \ \equiv \ \int_{0}^{\infty} u \, \exp \, (-sv) \, dv \tag{54}$$

Equation (52) may be solved exactly and providing the solution remains finite at all times, one finds

$$\zeta \ = \ c \, \exp \, (-y \sqrt{s}) + \zeta_i \, / s \tag{55}$$

After taking the Laplace transform of the boundary condition and substituting in Eq. (55), one finds

$$- \, c \, \sqrt{s} \, - \, (c + \zeta_i \, / \, s) \ = \ - \, \pounds \, \left(1 / u \, (0, v) \right) \tag{56}$$

and after solving the latter equation for c and substituting into Eq. (55), the expression for ζ becomes

$$\zeta \ = \ (\zeta_i \, / \, s) \, - \, \zeta_i \, \exp \, (-y \sqrt{s}) \, / \left[s \, (\sqrt{s} + 1) \right]$$

$$+ \ \left(\exp \, (-y \sqrt{s}) / (\sqrt{s} + 1) \right) \pounds \left(1 / u \, (0, v) \right) \tag{57}$$

Peters [18] shows that this latter equation may be inverted to give

$$u \, (y, v) \ = \ u \, (y, 0) \left(erf \, (y \, / 2 \sqrt{v}) + erfc \, (y \, / 2 \sqrt{v} + \sqrt{v}) \, \exp \, (y + v) \right)$$

$$+ \int_{0}^{v} \left(1 / u \, (0, W) \right) \left[\left(\exp \, (-y^2 / 4 \, (v - W)) \right) / \sqrt{\pi \, (v - W)} \right.$$

$$\left. - \, \exp \, (y + v - W) \, erfc \left(y \, / 2 \, \sqrt{v - W} + \sqrt{v - W} \right) \right] dW \tag{58}$$

At the boundary of the liquid phase

$$u \, (0, v) \ = \ u \, (0, 0) \, (erf \sqrt{v}) \, \exp \, (v)$$

$$\int_{0}^{v} \left(1 / u \, (0, W) \right) \left[\left(1 / \sqrt{\pi \, (v - W)} \right) - \exp \, (v - W) \, erfc \left(\sqrt{v - W} \right) \right] dW \tag{59}$$

We are now in a position to construct an approximate short-time solution that will be denoted $\tilde{u}$. We consider the case in which $u \, (y, 0)$ is zero. For short-times, i.e. for

123

$$v << 1 \tag{60}$$

and since

$$0 \leq W \leq v \tag{61}$$

the first term in the bracket of Eq. (59) will be large, i.e.

$$1/\sqrt{\pi(v - W)} >> 1 \tag{62}$$

whereas the second term will have unity as its maximum value. Thus for these times, we may approximate Eq. (59) and determine $\tilde{u}(0,v)$ from

$$\tilde{u}(0,v) = \int_0^v 1/u(0,W) \left[\left(1/\sqrt{\pi(v - W)}\right)\right] dW \tag{63}$$

We shall seek a solution of the form

$$\tilde{u}(0,v) = A v^B \tag{64}$$

where A and B are constants. After substituting Eq. (64) into Eq. (63), it is found that a solution of this form satisfies Eq. (63) provided that

$$B = 1/4 \tag{65}$$

and

$$A^2 = (2/\sqrt{\pi}) \int_0^{\pi/2} (\sin z)^{1/2} dz \tag{66}$$

or

$$A = 1.62736 \tag{67}$$

To construct the long-time solution, we shall use the step function approximation. The step function $\tilde{u}(v)$ is defined by the following:

$$\text{for } v_{n-1} < v < v_n \tag{68}$$

$$\tilde{u}_n = u(0, v_n) \tag{69}$$

From Eq. (59), $u(0, v_n)$ may be expressed

$$u(0, v_n) = u_i f(v_n) + \int_0^{v_n} \left(1/u(0,W)\, g(v_n - W)\right) dW \tag{70}$$

where u_i is $u(0,0)$ and

$$f(v) = erfc\sqrt{v} \left(\exp(v)\right) \tag{71}$$

and

$$g(z) = \left[1/\sqrt{\pi z} - \exp z\; erfc\sqrt{z}\right] \tag{72}$$

The interval is divided into steps. Equation (70) then becomes

$$u(0, v_n) = u_i f(v_n) + \sum_{i=2}^{n-1} \left[u(0, v_i)\right]^{-1} \int_{v_{i-1}}^{v_i} g(v_n - W)\, dW$$

$$+ (u_n)^{-1} \int_{v_{n-1}}^{v_n} g(v_n - W)\, dW \tag{73}$$

which may be written

$$\tilde{u}_n = u_i f(v_n) + \sum_{i=2}^{n-1} \left[u(0, v_i)\right]^{-1} \int_{v_{i-1}}^{v_i} g(v_n - W)\, dW$$

$$+ \tilde{u}_n^{-1} \int_{v_{n-1}}^{v_n} g(v_n - W)\, dW \tag{74}$$

or

$$(\tilde{u}_n)^2 - B_n \tilde{u}_n - I_n = 0 \tag{75}$$

where

$$B_n = u_i f(v_n) + \sum_{i=2}^{n-1} \left(u(0, v_i)^{-1}\right) \int_{v_{i-1}}^{v_i} g(v_n - W)\, dW \tag{76}$$

and

$$I_n = \int_{v_{n-1}}^{v_n} g\,(v_n - W)\,dW \tag{77}$$

The quadratic Eq. (75) may be solved and since only the positive solution is of interest

$$\tilde{u}_n = 0.5\left[B_n + (B_n^2 + 4I_n)^{1/2}\right] \tag{78}$$

This latter equation can be used to generate the solution of the problem stated in Eqs. (49) and (50), except for the initial period. An approach that has been shown to give an accurate solution is to use the short time solution up to v equal to 10^{-11}. The value calculated from the short-time solution, Eq. (64), at this value of v is then the starting value for the step solution. To calculate $\tilde{u}_2$ requires one application of Eq. (64), and to calculate $\tilde{u}_n$ require $n-1$ applications. The accuracy of this calculation procedure was shown by Peters [18] to be better than 0.1%.

The amount predicted to be absorbed when it is assumed that the interface is immediately saturated with the gas upon exposure is referred to as the "local equilibrium" solution and denoted as Q_{le}. To non-dimensionalize it, we multiply Eq. (7) by the factor $K_{IL} / \alpha^2 D$, and when the liquid phase is initially free of dissolved gas write

$$Q'_{le} = 2\sqrt{v / \pi} \tag{79}$$

where v is given by Eq. (47).

The corresponding non-dimensional amount predicted to be absorbed when the kinetic expression for the rate of absorption (Eq.(43)) is used as the boundary condition, Q'_{ne}, can similarly be obtained from Eq. (51). One finds

$$Q'_{ne} = \int_o^v \left[\frac{1}{u\,(0,\,v')} - u\,(0,\,v')\right] dv \tag{80}$$

A comparison may now be made between these two predictions of the amount absorbed. The results obtained are shown in Fig. 2. As may be seen there, for small non-dimensional exposure times, there is a very large difference between the two predictions of the amount absorbed. However, it should also be noted that the non-dimensional time v depends on the liquid-gas combination through its dependence on α, D and K_{IL}. Thus a small exposure time for a particular case, such as oxygen and water, may not be a small exposure time for another liquid-gas combination. In the following section, we propose to investigate the possibility that the difference in the properties of the O_2 - H_2O (l) and CO_2 - H_2O (l) systems means that for the same exposure time, say on the order of 1 millisecond, such as that obtainable with the liquid-jet technique, the non-dimensional exposure time is smaller than unity for the O_2 - H_2O (l) system, but it is larger than unity for CO_2 - H_2 (l) system. If this is true, it would provide an explanation for why the liquid-jet absorption data gives an unreasonably small value for the diffusion coefficient for O_2 - H_2 (l) but the same experimental technique leads to a reasonable value for CO_2 - H_2O (l).

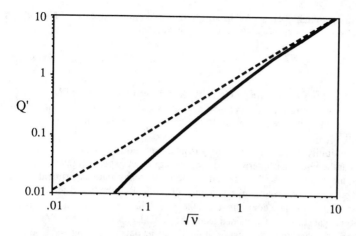

FIGURE 2. Dimensionless plot of the amount of gas absorbed as a function of the non-dimensional exposure time for two different assumed conditions at the liquid-gas interphase: 1) local equilibrium (dashed line), and 2) the kinetic expression for the rate of gas absorption that is obtained from Statistical Rate Theory (solid line). Note that if v is larger than unity, the local equilibrium assumption is valid, but if v is smaller than unity, this assumption leads to a poor prediction of the non-dimensional amount absorbed, Q'.

APPLICATION OF RATE EXPRESSION

Using the liquid-jet technique, Chiang and Toor [8] exposed water to O_2 at 22.2°C for periods of time ranging from 0.8 to 11.8 ms and, using the local equilibrium approximation at the interphase, calculated a diffusion coefficient of 1.6×10^{-5} cm^2/s.

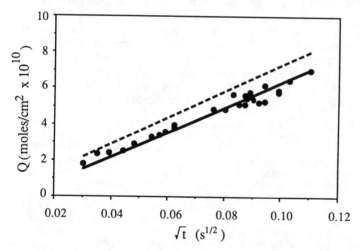

FIGURE 3. Measured (•) and predicted gas absorption versus time. The measured absorption is from Ref. [8] for O_2 in water at 22.2°C. The predicted amount of gas absorption assuming local equilibrium at the liquid-gas interphase is shown as the dashed line and that predicted from the kinetic expression obtained from Statistical Rate Theory for the rate of gas absorption at the interphase is shown as the solid line.

To compare this value with that obtained by the steady state experiment at 25°C of Ng and Walkley [9] that was conducted at 25°C, we apply the Einstein relation

$$D = \frac{CT}{\mu} \tag{81}$$

where C is a constant and μ is the kinematic viscosity. The extrapolated value of D obtained from the steady state experiment is found to be 2.48 x 10^{-5} at 22.2°C. Clearly the values obtained from the two different experimental techniques are inconsistent.

To investigate the possibility that this inconsistency results from assuming local equilibrium to exist at the liquid-gas interphase during the liquid-jet experiment, we shall assume the value of the diffusion coefficient obtained from the steady state experiment of Ng and Walkley to be the true value and investigate the fit obtained of the liquid-jet data using this true value of the diffusion coefficient and the expression for the rate of gas absorption at the liquid-gas interphase obtained using the Statistical Rate Theory. We first determine the value of K_{IL} that gives the best fit of the liquid-jet data of Chiang and Toor.

The fit of the Chiang and Toor data that is obtained by this procedure is shown in Fig. 3. The conditions used in the calculations are listed in Table 1, along with the value of the radius of the absorption cross-section obtained by this procedure. This value (1.80 x 10^{-8} cm) is in very close agreement with the values of the molecular radius of O_2 tabulated in Tabor [19] based on three different methods of determination (average value of 1.78 x 10^{-8} cm).

Now we investigate the consistency of this approach by investigating whether the assumption of local equilibrium is valid for the steady state experiment of Ng and Walkley. This may be done by considering the value of v for the O_2 - H_2O system using the values listed in Table 1 and Eq. (47), one finds that v has a value greater than 10 whenever the exposure time is greater than 0.015 s. As may be seen from Fig. 2, whenever v is greater than 10, there is no essential difference between the amount of gas predicted to be absorbed on the basis of the local equilibrium assumption and when the Statistical Rate Theory approximation is used. Since the exposure time in the Ng and Walkley experiments was on the order of 3000s, the local equilibrium assumption would be justified during 99.9995% of their experimental period. Hence the value of the diffusion coefficient inferred from their experimental data should be valid.

Clearly then, in the case of O_2 - H_2O (l), taking into account the rate of gas absorption at the interphase during the absorption process provides an explanation for the "surface resistance" discussed by Chiang and Toor. We now turn to the question of whether the same approach gives an explanation for why no discrepancy is found when the local equilibrium approximation is used to infer the value of the diffusion coefficient for CO_2 in H_2O when the same experimental techniques are used to measure the amount of CO_2 absorbed. Of the various factors relevant to gas absorption (see Eqs. (36) and (51)), CO_2 differs from O_2 in three important aspects; i) CO_2 is approximately 29 times more soluble in water at 22.2°C than is O_2, ii) CO_2 is approximately 75% as diffusive as O_2, and iii) CO_2 has a larger molecular radius as determined from the hard sphere model of kinetic theory and measured values of the viscosity and thermal diffusivity [19].

TABLE 1. Values used for calculations of gas absorption in a liquid-jet of water at 22.2°C.

	O_2	CO_2
P	1.01325×10^6	1.01325×10^6 dyn/cm^2
P_∞	2.676×10^4	2.676×10^4 dyn/cm^2
c_1	5.5384×10^{-2}	5.5384×10^{-2} $moles/cm^3$
k_H	4.1980×10^{10}	1.4476×10^9 dyn/cm^2
D	2.48×10^{-5}	1.85×10^{-5} cm^2/s
$\sigma_2{}^\dagger$	1.80×10^{-8}	2.59×10^{-8} cm
K_{IL}	1.7299×10^{-7}	1.2744×10^{-5} $moles/cm^2s$

† inferred value using Statistical Rate Theory

However, the molecular radius estimated from the value of kinetic properties arising from the interaction of CO_2 molecules with other CO_2 molecules cannot be expected to give an accurate estimate of the molecular radius when CO_2 molecules are interacting with H_2O molecules because of the chemical reaction that is known to occur between these latter two components [15].

We proceed in the same manner to examine the gas absorption data for CO_2 absorbing into H_2O (l) as we did for O_2 dissolving into H_2O (l). Namely, we take as valid the value of the diffusion coefficient for CO_2 - H_2O as determined from the steady state method of Ng and Walkley [9], and when their measurement is extrapolated to 22.2°C using Eq.(81), one finds a value of 1.85×10^{-5} cm^2/s. Using this value of D, we determine the value of K_{GI} which is the best fit of the experimental data obtained from the liquid-jet technique by Raimondi and Toor [13]. The fit of their data obtained from this procedure is shown in Fig. 4.

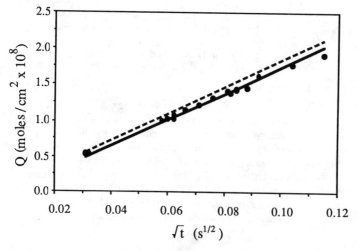

FIGURE 4. Measured ($\bullet$) and predicted gas absorption versus time. The measured absorption is from Ref. [13] for CO_2 in water at 22.2°C. The predicted amount of gas absorption assuming local equilibrium at the liquid-gas interphase is shown as the dashed line and that predicted from the kinetic expression obtained from Statistical Rate Theory for the rate of gas absorption at the interphase is shown as the solid line.

The value of K_{IL} that gives this fit is listed in Table 1. From Eq. (36) and this value of K_{IL}, the value of the absorption cross-sectional radius has been inferred. Its value is given and also listed in Table 1. The average value of the molecular radius obtained from the hard sphere model of kinetic theory and measured values of the viscosity, thermal conductivity and self-diffusion coefficient is 2.13×10^{-8} cm. This value is approximately 18% smaller than the absorption cross-sectional radius inferred from the liquid-jet data. However, this difference may be physically correct in this case because, unlike O_2 absorbing in H_2O, there is a chemical reaction between CO_2 and H_2O. If one examines the derivation of K_{IL}, the absorption cross-sectional radius is used to calculate the cross-sectional area of a number of absorbed CO_2 molecules. These molecules would be interacting primarily with H_2O molecules; thus because of the known reaction between CO_2 and H_2O, the electron structure of the absorbed CO_2 molecules would not be expected to be confined to the same extent as when CO_2 molecules are colliding with other CO_2 molecules, as described in kinetic theory and used to infer the smaller molecular radius. This explanation for the inferred value of the absorption cross-sectional radius being larger than the kinetic theory value must be viewed as speculation at this time. However, we note that the theory could be examined further by carrying out a liquid-jet study with another nonreacting gas.

If one calculates v for the CO_2 - H_2O (l) system, using the values in Table 1 and Eq. (47), one finds that it is greater than 10 for exposure time of greater than 1.7 ms. The experimental procedure of Ng and Walkley involved exposure times that were six orders of magnitude greater than this. Thus according to the Statistical Rate Theory approximation, the local equilibrium assumption would be valid in their case for this system as well.

SUMMARY

If a liquid phase that is initially free of a dissolved gas is suddenly exposed to a gas at the phase boundary, previous studies have usually assumed that local equilibrium was established immediately when the liquid phase was exposed to the gas. This has led to the illogical consequence that the value of the diffusion coefficient inferred from experiments involving absorption depended on the experimental technique used. In the instance of O_2 absorption in water at 22.2°C, Chiang and Toor [8] reported a value of 1.6×10^{-5} cm^2/s using a liquid-jet technique, whereas Ng and Walkley [9] reported a diffusion coefficient of 2.48×10^{-5} cm^2/s (extrapolated value) using a steady state bubble method. The low value reported by Chiang and Toor was reluctantly attributed to "surface resistance" since results obtained for CO_2 absorption in water using the same two techniques [13] were found to be in agreement.

Statistical Rate Theory allows the equilibrium assumption to be relaxed. It uses the concept of transition probabilities and introduces an assumption regarding the net rate of transition between possible microscopic states. When the expression for the rate of gas absorption that is obtained from this procedure is used as the boundary condition for the diffusion equation, the amount of gas absorption as a function of time can be predicted. The results show that if the parameter v (i.e. $t K_{IL}^2 / \alpha^2 D$) is small compared to unity, then the nonequilibrium rate of gas absorption plays a major role. Whereas if the parameter v is larger than unity, there is little difference in the predicted amount of absorption whether it is assumed that local equilibrium exists at the phase boundary or the rate of gas absorption at the phase boundary is taken into account.

If the value of v is calculated for O_2 dissolving into H_2O at 22.2°C using the value of D obtained from steady state measurements, one finds that v has a minimum value of 0.623 for the liquid-jet measurements of Chiang and Toor [8]. If one makes the same calculation for CO_2 dissolving in H_2O, then the smallest value v for the liquid-jet measurements of

Raimondi and Toor [13] is found to be 5.4. Therefore, in the case of O_2 absorbing into H_2O as measured by the liquid-jet technique, it is important to take into account the rate of gas absorption. Whereas for CO_2, the interphase goes to equilibrium so fast that taking into account the rate of gas absorption for the liquid-jet measurements is not important. Note that this difference in the rate of O_2 or CO_2 absorption into H_2O arises from their physical properties.

REFERENCES

1. Ward, C.A., Findlay, R.D. and Rizk, M. Statistical Rate Theory of Interfacial Transport. I. Theoretical Development, *J. Chem. Phys.*, vol. 76, no. 11, pp. 5599–5605, 1982.

2. Ward, C.A, Rizk, M. and Tucker, A.S. Statistical Rate Theory of Interfacial Transport. II. Rate of Isothermal Bubble Evolution in a Liquid-Gas Solution, *J. Chem. Phys.*, vol. 76, no. 11, pp. 5606–5614, 1982.

3. Ward, C.A., The Rate of Gas Absorption at a Liquid Interface., *J. Chem. Phys.*, vol. 67, no. 1, pp.229–235, 1977.

4. Ward, C.A. and Elmoselhi, M. Molecular Adsorption at a Well Defined Gas-Solid Interphase: Statistical Rate Theory Approach, *Surf. Sci.*,vol. 176, pp. 457–475, 1986.

5. Suresh, A.K., Gupta, R.K. and Sridhar, T., *Frontiers in Chem. Reaction Engin.*, eds. L.K. Doraiswamy and R.A. Mashelkar, vol. 1, pp. 489–498, Wiley Eastern, New Delhi, 1984.

6. Danckwerts, P.V. *Gas-Liquid Reactions,* pp. 66–69, McGraw-Hill, New York, 1970.

7. Danckwerts, P.V. and Kennedy, A.M. The Kinetics of Absorption of Carbon Dioxide into Neutral and Alkaline Solutions, *Chemical Engineering Science, vol. 8, p. 201, 1958.*

8. Chiang, S.H. and Toor, H.L. Interfacial Resistance in the Absorption of Oxygen by Water, *A.I. Ch.E. Journal,* vol. 5, no. 2, pp. 165–168, 1959.

9. Ng, W.Y. and Walkley, J. Diffusion of Gases in Liquids: The Constant Size Bubble Method, *Cdn. J. Chem.*, vol. 47, pp. 1075–1077, 1969.

10. Higbie, R. The Rate of Absorption of a Pure Gas Into a Still Liquid During Short Periods of Exposure, *A. Instit. Chem. Engineers,*vol. 31, pp. 365–388, 1935.

11. Danckwerts, P.V. and Kennedy, A.M. Kinetics of Liquid-Film Processes in Gas Absorption. Part II: Measurements of Transient Absorption Rates. *Trans. Instn. Chem. Engrs.* (London), vol. 32, pp. S53, 1954.

12. Danckwerts, P.V. *Insights into Chemical Engineering,* pp. 1–3, Pergamon, Oxford, 1981.

13. Raimondi, P. and Toor, H.L. Interfacial Resistance in Gas Absorption, *A.I. Ch.E. Journal,* vol.5, no. 4, pp. 86–92, 1959.

14. Ward, C.A. Effect of Concentration on the Rate of Chemical Reactions, *J. Chem. Phys.*, vol. 5, no. 11, 5605–5615, 1983.

15.   Cotton, F.A. and Wilkinson, G. *Advanced Inorganic Chemistry,* pp. 36, Wiley, Toronto, 1980.

16.   Tolman, R. C. *Principles of Statistical Mechanics,* pp.395, Oxford University, London, 1938.

17.   Landau, L. and Lifschitz, *E.M. Statistical Physics,* pp. 351, Addison Wesley, New York, 1958.

18.   Peters, J. Solution of Diffusion Problem Associated with a Nonlinear Boundary Condition at a Liquid-Gas Interface. M. Appl. Sc. Thesis, University, Toronto, 1972.

19.   Tabor, D., *Gases, Liquids and Solids,* pp. 72, Cambridge University Press, Cambridge, 1979.

Chapter 6

Shapes of Fluid Particles

JOHN R. GRACE
Department of Chemical Engineering
University of British Columbia
2216 Main Mall
Vancouver, V6T 1W5 Canada

TOM WAIREGI
Marathon Oil Company
539 South Main Street
Findlay, Ohio 45840, USA

INTRODUCTION

Liquid drops and gas bubbles together constitute fluid particles. They occur in many natural phenomena (e.g. rain, boiling) and in a host of industrial processes (e.g. liquid extraction, spray drying, fermentation). The shape of the fluid particle affects its rising or falling velocity, and hence its residence time in the field or apparatus of interest. Moreover, heat and mass transfer between the fluid particles and the surrounding medium (usually called the "continuous phase") are greatly influenced by the interfacial area, which in turn is determined by the volume of the fluid particle and by its shape.

In this chapter we consider the factors which influence the shape of gas bubbles rising through liquids, of liquid drops falling through gases, and of liquid drops rising or falling through immiscible liquids. For simplicity, we ignore interactions between fluid particles, and we treat only fluids which are Newtonian, not accelerating (i.e. static or travelling at their terminal rising or falling velocity). The continuous phase is assumed to be stagnant remote from the fluid particle of interest. A method is given of predicting the shapes of bubbles and drops which are in free, steady motion in an immiscible fluid of large extent. This method explains the deformations which occur as bubbles and drops grow in size.

STATIC DROPS AND BUBBLES

Consider the case of a fluid particle which is constrained to be stationary by a flat plate blocking its path, causing it to be a "sessile" drop or bubble as shown in Fig. 1(a) and (b). Alternatively, the bubble or drop may be held back against the "pull" of gravity by capillary forces, as with the familiar cases of drops forming at a dripping faucet or intravenous drip chamber. It is then said to be "pendant", and typical shapes are illustrated in Fig. 1(c) and (d).

The shapes of sessile or pendant fluid particles have long been used to provide a measure of the interfacial or surface tension, σ, between two fluid phases (i.e. between two liquids or between a gas and a liquid). Under static conditions the drop or bubble must have a shape for which the normal stresses on either side of the interface are counterbalanced. Interfacial tension forces act to minimize surface energy, hence tending to minimize distortion from the spherical. Hence the shape is a function both of the volume of the static fluid particle and of the interfacial or surface tension.

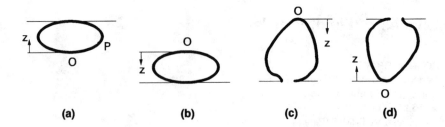

FIGURE 1. Shapes of static drops and bubbles: (a) sessile drop or bubble ($\rho' < \rho$); (b) sessile drop ($\rho' > \rho$); (c) pendant drop or bubble ($\rho' < \rho$); (d) pendant drop ($\rho' > \rho$).

Consider the sessile bubble or drop shown in Figure 1(a). A balance of normal stresses at the nose (point 0) yields

$$p'_0 = p_0 + 2\sigma/R_0 \tag{1}$$

where the prime denotes quantities on the inside of the fluid particle, while the unprimed quantities refer to the continuous (or outer) phase. Here R_0 is the radius of curvature at the nose. A similar balance of normal stresses at a general point, P, on the surface yields

$$p' = p + \sigma(1/R_1 + 1/R_2) \tag{2}$$

where R_1 and R_2 are principal radii of curvature at point P. Since both fluids are static, hydrostatic pressure distributions must apply on both sides of the interface, i.e.

$$p = p_0 - \rho gz \tag{3}$$

and

$$p' = p'_0 - \rho'gz \tag{4}$$

where z is a coordinate measured vertically from point 0 in the direction of the fluid particle as shown in Fig. 1. Substituting equations (3), (4) and (1) into equation (2), it is straightforward to show that

$$- (\rho - \rho')gz = \sigma(2/R_0 - 1/R_1 - 1/R_2) \tag{5}$$

Similar analyses can be performed for cases (b), (c) and (d) in Figure 1 leading to the general result

$$\pm \Delta\rho gz = \sigma(2/R_0 - 1/R_1 - 1/R_2) \tag{6}$$

where $\Delta\rho$ is the magnitude of the difference in density between the two fluids, and where the positive and negative signs on the left hand side refer to the pendant and sessile cases, respectively. This analysis can easily be extended to other cases, e.g. where a drop floats at the interface between two immiscible fluids or stationary fluid particles subject to applied forces.

This equation shows immediately why static drops and bubbles deform in the observed manner. For cases (c) and (d) in Figure 1, the left-hand side must be positive since we have a pendant drop and a positive value of z. Hence $1/R_1 + 1/R_2$ must be less than $2/R_0$, meaning that the radius of curvature must be a minimum at point 0. Thus the drop must elongate. Conversely, for cases (a) and (b), the left side is negative, requiring that the radius of curvature be a maximum at the nose (point 0) of the drop or bubble, so that the shape must become flattened.

When substitutions are made for R_1 and R_2, a second order ordinary differential equation results that must be solved numerically (Bashforth and Adams, 1883). Accurate tabulations of resultant shapes, as well as a general review, approximate solutions and numerical methods, are given by Hartland and Hartley (1976).

DROPS AND BUBBLES IN FREE MOTION

When unconstrained fluid particles rise or fall due to gravity forces, there are changes in the pressure fields due to the motion, as required by the Navier-Stokes equation for the dispersed fluid. Inertial forces can lead to other types of deformation from those encountered with static fluid particles.

It is convenient to treat separately low and high Reynolds number cases, and then to consider empirical methods for predicting the shapes of steadily rising or falling drops and bubbles.

Low Renolds Number

In the limiting case of pure (surfactant-free) fluids such that both

$$Re = \rho d_e v_T/\mu \rightarrow 0 \tag{7}$$

and

$$Re' = \rho' d_e v_T/\mu \rightarrow 0 \tag{8}$$

where d_e is the volume-equivalent sphere diameter and v_T the terminal (rising or falling) velocity, then the motion of the fluid particles is described by the well-known Hadamard-Rybczynski solution (Hadamard, 1911; Ribczynski, 1911). The shape of the fluid particle is spherical, with the internal and external Stokes stream functions given, respectively, by

$$\psi' = \frac{v_T r^2 \sin^2 \theta}{4(\kappa + 1)} [1 - r^2/a^2] \tag{9}$$

and

$$\psi = \frac{-v_T r^2 \sin^2 \theta}{2} \left[1 - \frac{a(3\kappa + 2)}{2r(\kappa + 1)} + \frac{\kappa a^3}{2r^3 (\kappa + 1)} \right] \tag{10}$$

where a is the radius of the fluid sphere and $\kappa = \mu' / \mu$ is the ratio of the viscosities of the internal and external fluids. The terminal velocity can be shown to be given by

$$v_T = \frac{2g \, \Delta\rho \, a^2}{3\mu} \frac{(\kappa + 1)}{(3\kappa + 2)} \tag{11}$$

This solution completely satisfies all boundary conditions, including the normal stress conditions, i.e. a spherical fluid particle shape is appropriate, even in the absence of surface or interfacial tension forces, providing that the above assumptions (very low Re and Re', absence of surface contaminants) are valid.

Let us now consider what happens as we begin to relax the above assumptions one by one. First consider the case where Re is still small, but Re' is no longer constrained to be small. In this case (Pan and Acrivos, 1968), fluid particle shapes are predicted to differ very little from the case where Re and Re' are both small, even if Re' >> 1.

Secondly, consider the case where surface active contaminants begin to inhibit the internal motion of the fluid particle. For small amounts of contamination, the observed effect is for contaminants to be swept to the rear where they accumulate, causing a "stagnant cap" at the rear (Savic, 1953; Garner and Skelland, 1955). The resulting internal circulation pattern is then distorted so that there is no longer fore-and-aft symmetry. Various analyses have been presented to account for the changes in internal motion (e.g. Savic, 1953; Davis and Acrivos, 1966; Huang, 1968) caused by a stagnant cap. The key point, as far as the present review of particle shapes is concerned, is that the modified flow and lack of fore-and-aft symmetry due to accumulation of surface active impurities do not, in themselves, cause appreciable deformation from the spherical shape for practical cases.

The findings summarized so far explain why the simple regime diagram introduced by Grace and coworkers (Grace, 1973; Grace et al, 1976; Clift et al., 1978) and modified by Bhaga and Weber (1981) shows that fluid particles remain in the "spherical regime" if Re is sufficiently small, regardless of the surface tension, Re' or surface contamination.

For freely moving fluid particles, the major deformation which occurs is due to inertial effects in the external fluid. For small deformation, Taylor and Acrivos (1964) used the method of matched asymptotic expansions to obtain, to terms of order We^2/Re,

$$\frac{r(\theta)}{a_0} = 1 - \lambda \, We \, P_2 (\cos \theta) - \frac{3\lambda(11\kappa + 10)}{70(\kappa + 1)} \frac{We^2}{Re} P_3 (\cos \theta) \tag{12}$$

where a_0 is the radius of the undeformed sphere, P_2 and P_3 are second and third order Legendre polynomials, We is the Weber number given by

$$We = 2\rho\, a_0\, v_T^2 / \sigma \qquad (13)$$

and

$$\lambda = \frac{[\,3 + 10.3\kappa + 11.4\kappa^2 + 4.05\kappa^3\,] - (\kappa + 1)\,(\rho' - \rho)\,/\,3\rho)}{32(\kappa + 1)^3} \qquad (14)$$

The above results were for small Re', but were extended to larger Re' by Pan and Acrivos (1968). For small but non-zero values of We, the fluid particle is predicted to deform to be an exact spheroid which, for properties of physical interest, is oblate. Equation (12) has been extended to greater number of terms by Brignell (1973) and by Akiyama and Yamaguchi (1975).

For all but a small number of very viscous liquids where

$$M = g\mu^4 \Delta\rho / \rho^2 \sigma^3 \gg 1 \qquad (15)$$

interfacial or surface tension forces are large enough to keep deformation small until Re is relatively large. Hence, while the above approach is useful in showing that inertial forces associated with deviations from very low Re can lead to deformation into an oblate spheroidal profile, it is of limited practical application. We turn therefore to deformations which take place at higher Re, i.e. for systems with $M \ll 1$.

High Reynolds Number

A simple approach can be used to illustrate why fluid particles deform in the observed manner when they are in steady unconstrained motion at high Re. The shape of fluid particles under such conditions is primarily a function of hydrostatic pressure, interfacial or surface tension, and external hydrodynamic pressure forces. Other forces, such as those due to internal circulation or electrostatic forces, are of secondary importance. A full numerical solution, with the shape of the fluid particle unknown, would be required to fully resolve the problem. However, a simple procedure, first proposed by Savic (1953) and later extended by Pruppacher and Pitter (1971) for water drops in air and by Wairegi (1974) for fluid particles in liquids, provides useful approximate results. In this case, the shape of the fluid particle was described in terms of Legendre polynomials or Fourier series. Shapes predicted for water drops falling in air are shown in Figure 2. Note that for small drops the profile is nearly spherical, but as the drop size increases, both the front (lower edge) and rear (top edge) are flattened, with greater flattening occurring at the lower edge. These trends are in general agreement with experimental observed shapes for water drops in air (Pruppacher and Beard, 1970). However, the predicted shapes show the development of an indentation or dimple along the leading edge for drops of $d_e > 4$ mm, something which is never observed in practice for steady fall conditions.

In the above model, it was assumed that the aerodynamic pressure distribution on the drop is identical to that for a rigid sphere at the same Re. In practice, of course, the pressure distribution will vary considerably as the fluid

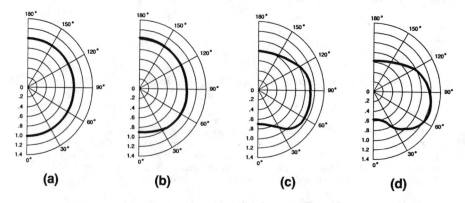

(a)   **(b)**   **(c)**   **(d)**

FIGURE 2.   Shapes of water drops falling freely in air at 20C and 101.3 kPa
             predicted by Wairegi (1974) using a method similar to Savic (1953)
             but with the shape represented using Legendre polynomials
             rather than Fourier series.   Aerodynamic pressure distribution is
             assumed to be the same as for a rigid sphere at the same   Re.
             (a) d_e = 0.25 mm;   (b) d_e = 0.86 mm;   (c) d_e = 3.5 mm   (d) d_e = 7.4 mm.
             See Figure 4 for corresponding actual shapes.

particle deforms.   To be able to predict substantial deformation, it is therefore
essential to be able to adjust the pressure distribution to conform to the
modifications in the shape of the fluid particle.   Here we describe a simple
method, called the double semi-ellipsoid technique, for predicting, to a first
approximation, the shape of drops and bubbles, at high   Re.   The following
assumptions are employed:

(a) As in the approach employed by Finlay (1957), both the leading and
    trailing edges of the fluid particle are represented by semi-ellipsoids of
    revolution, both oblate, having a common (vertical) axis of symmetry and
    semi-major axes, as shown in Figure 3.
(b) Any secondary motion, such as shape dilations, wobbling, or spiral motion
    (Clift et al., 1978) is ignored.

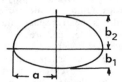

FIGURE 3.   Double semi-ellipsoid shape, with b_1 and b_2 to be determined
             based on given value of   a   and fluid properties.

(c) The hydrodynamic pressure distribution due to the flow over the leading semi-ellipsoid is assumed to be identical to that for potential flow over a complete oblate spheroid of the same eccentricity and orientation as the semi-spheroidal forward portion.

(d) It is assumed that flow separation occurs at the "equator" (largest cross-section) and that there is negligible pressure recovery over the aft portion of the fluid particle.

These approximations are clearly crude. While the potential flow pressure distribution is commonly a good approximation to real pressure distributions near the leading edge, especially for streamlined shapes and for low κ with freely circulating inner fluid, this approximation is not expected to be very accurate near the equator, as the front portion flattens more and more, or for highly viscous inner fluids. Similarly, some pressure recovery is expected to occur in practice for the rear portion, the extent linked to the degree of flattening and the viscosity ratio, κ. Despite the approximate nature of assumptions (c) and (d), we will see that the approach gives good qualitative predictions and illustrates the factors which affect the shape of fluid particles.

With these approximations, the analysis proceeds as follows:

(1) <u>Anterior semi-spheroid</u>. The dynamic pressure around the anterior semi-spheroid is obtained by applying Bernoulli's equation between the nose ($\eta_1 = 0$) and a general point on the surface, giving

$$p_{dyn} = p_{01} + \frac{1}{2} \rho v_T^2 \ (1 - q^2/v_T^2) \tag{16}$$

where q is the magnitude of the fluid velocity relative to axes fixed on the drop or bubble. A balance of dynamic, hydrostatic and interfacial or surface tension forces at a general point on the surface then yields

$$\sigma\left[\frac{1}{R_1} + \frac{1}{R_2}\right] + \frac{1}{2}\rho v_T^2\left(1 - \frac{q^2}{v_T^2}\right) - \rho g b_1 \cos \eta_1 = p_{01} \tag{17}$$

At the nose, i.e. at $\eta_1 = 0$, eqn. (17) reduces to

$$2\sigma/R_0 + \frac{1}{2}\rho v_T^2 - \Delta\rho g b_1 = p_{01} \tag{18}$$

Noting that $R_0 = a^2/b_1$ and subtracting eqn. (17) from eqn. (18), we obtain

$$\sigma\left[\frac{2b_1}{a^2} - \left(\frac{1}{R_1} + \frac{1}{R_2}\right)\right] + \frac{1}{2}\rho q^2 = \Delta\rho g\, b_1(1 - \cos\eta_1) \tag{19}$$

For an oblate spheroid (Saffman, 1956) we may write

$$\frac{1}{R_1} + \frac{1}{R_2} = \frac{1}{a}\frac{(1 - e_1^2)^{1/2}\ (2 - e_1^2 \sin^2 \eta_1)}{(1 - e_1^2 \sin^2 \eta_1)^{3/2}} \tag{20}$$

where

$$e_1^2 = 1 - b_1^2/a^2 \tag{21}$$

The tangential velocity at the surface of the spheroidal surface for potential flow is given by

$$q^2 = \frac{v_T^2 \sin^2 \eta_1}{K_3^2 [(1 - e_1^2) \sin^2 \eta_1 + \cos^2 \eta_1]} \tag{22}$$

where

$$K_3 = \frac{1}{e_1^3} (\sin^{-1} e_1 - e_1 \sqrt{1 - e_1^2}) \tag{23}$$

With these substitutions and converting to dimensionless form, we may finally write eqn. (19) as

$$\frac{Eo^*}{4} \sqrt{1 - e_1^2} \, (1 - \cos \eta_1) = \sqrt{1 - e_1^2} \left[2 - \frac{(2 - e_1^2 \sin^2 \eta_1)}{(1 - e_1^2 \sin^2 \eta_1)^{3/2}} \right]$$
$$+ \frac{We^* \sin^2 \eta_1}{4K_3^2 (1 - e_1^2 \sin^2 \eta_1)} \tag{24}$$

where the three terms arise from hydrostatic, capillary and dynamic terms, respectively.

Note that both an Eötvös number and a Weber number

$$Eo^* = \frac{\Delta \rho g \, (2a)^2}{\sigma} ; \quad We^* = \frac{\rho \, (2a) \, v_T^2}{\sigma} \tag{25}$$

turn out to be important. Equation (24) may be examined for limiting cases. When Eo^* and We^* both approach 0, i.e. σ is very large, then e_1 approaches 0, i.e. the drop or bubble is spherical as expected. At the other extreme, for small σ, the middle term can be neglected; the resulting equation cannot be satisfied for all η_1, it leads to the appropriate expression for the terminal velocities of oblate spheroidal-spherical-cap bubbles or drops (Davies and Taylor, 1950).

For the more general case, for given Eo^* and We^*, appropriate values of e_1 were found, using a Golden section search technique, which most nearly satisfied equation (24) in the sense that the sums of squares of the diviations for $\eta_1 = 0.1, 0.2, \ldots 0.8$ were minimized. With e_1 thus found, the semi-minor axis, b_1, is then found from equation (21).

(2) Posterior semi-spheroid. From assumption (d) given above, it is clear that the rear portion of the fluid particle should deform exactly like a sessile drop. A simple estimate of b_2, the semi-minor axis of the posterior semi-ellipsoid, can be obtained by matching pressures at the equator and at the rear. Starting with equation (6) (with the negative sign on the left-hand

side), substituting $R_0 = a^2/b_2$, $R_1 = b_2^2/a$, $R_2 = a$ and $z = b_2$, it is easily shown that (b_2/a) must satisfy the cubic equation

$$(8 + Eo^*) (b_2/a)^3 - 4(b_2/a)^2 - 4 = 0 \qquad (26)$$

Hence (b_2/a) can be estimated by solving equation (26). Alternatively, conventional results for sessile drops (Hartland and Hartley, 1976) may be used. However, given the approximate nature of the assumptions, equation (26) was used here to estimate the semi-minor-axis, and hence the shape of the posterior semi-spheroid.

(3) Predictions. This method provides estimates of the semi-minor axes, b_1 and b_2, for a given (common) semi-major axis, a, given the fluid properties and a measured or predicted value of the terminal rising or falling velocity, v_T. One can easily calculate the volume-equivalent diameter, d_e, by summing the volumes of the two semi-ellipsoids so that

$$d_e = [4a^2 (b_1 + b_2)]^{1/3} \qquad (27)$$

Note that there are no fitted parameters in the resulting shape predictions.

Predictions have been made (Wairegi, 1974) for a wide range of experimental fluid-fluid systems where there are detailed experimental results available in the literature and where M is sufficiently small that deformation occurs at high Re. The method predicts that there is a gradual and progressive flattening of the posterior portion of the fluid particle, whatever the fluid-fluid system. However, the anterior portion is more complex. In this case, interfacial or surface tension forces always act to maintain a spherical shape, whereas dynamic forces act to flatten the front, while hydrostatic forces tend to elongate the front, as for a pendant drop. Depending on the relative magnitudes of these three, shape changes at the front may occur more or less quickly than at the rear of the fluid particle.

Consider the familiar case of water drops falling in air. Shapes obtained from experimental work by Pruppacher and Beard (1970) are compared with predicted shapes in Figure 4. Both the experimental results and the

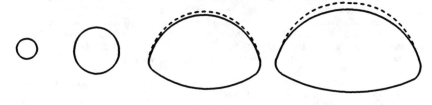

(a)   **(b)**   **(c)**   **(d)**

FIGURE 4. Shapes of water drops falling freely in air at 20C and 101.3 kPa. Solid lines give predictions from double ellipsoid method, while broken lines give corresponding actual shapes, where they differ appreciably, obtained from Pruppacher and Beard (1970). (a) d_e = 0.25 mm; (b) d_e = 0.86 mm; (c) d_e = 3.5 mm; (d) d_e = 7.4 mm.

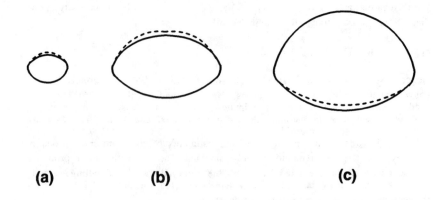

(a)   **(b)**   **(c)**

FIGURE 5. Shapes of air bubbles rising freely through water at 20C. Solid lines give predictions from double ellipsoid method; broken lines give corresponding experimental shapes (Wairegi, 1974). (a) d_e = 5.4 mm; (b) d_e = 8.1 mm; (c) d_e = 13.6 mm.

predictions show that the shape is initially closely spherical, but flattening occurs at both front (bottom) and back (top), especially at the former, as the drop size increases. While the qualitative results are correct, the predictions tend to show somewhat more deformation than is experienced in practive, especially on the posterior surface.

Predicted and experimental results for air bubbles rising in water are shown in Figure 5. This case is especially interesting because experimental results (e.g. Haberman and Morton, 1953) show that more flattening initially occurs at the front (top) than at the back (bottom), but that with increasing bubble volume, (transition at $d_e \doteq 8.5$ mm) this trend later reverses and the shape eventually approaches (for $d_e \gtrsim 18$ mm) that of a spherical-cap, with severe flattening at the rear and a nearly spherical anterior shape. It is seen that the simple theory gives qualitatively the correct result, although there are again quantitative differences. In the experimental case, bubbles of intermediate sizes undergo secondary motion (shape dilations of a periodic nature combined with wobbling) which not only make comparisons difficult, but also violate one of the assumptions of the model. Nevertheless, it is encouraging that correct qualitative trends are so well predicted.

Comparisons have been made with a wide range of other experimental systems, in particular for drops of various different liquids in air (Finlay, 1957), various liquid-liquid systems (Wellek et al, 1966) and bubbles and drops in aqueous sucrose solutions, ethylene glycol and paraffin oil (Wairegi, 1974). In almost every case, qualitative results, in terms of overall shapes and whether flattening occurs primarily at the front or rear, were reasonable from the simple theoretical method.

We conclude that the simple method gives a useful means of explaining the nature of the shape changes that occur at high Re for drops and bubbles rising or falling freely through immiscible liquids or gases. The key point is that the method correctly accounts for systems in which more flattening occurs at the front, at the back, or first at one and then the other. Most previous simple methods for predicting the shapes of fluid particles (e.g. Green, 1975) have assumed fore-and-aft symmetry, an assumption which is

142

clearly seriously in error. No doubt quantitive predictions could be improved, e.g. by incorporating empirical constants or using improved methods of predicting the dynamic pressure distributions, but this may not be warranted given the success of empirical methods for predicting key shape parameters (e.g. see Clift et al, 1978).

Empirical Relationships

It is commonly sufficient to represent fluid particles as falling within one of three principal shape regimes - spherical, "ellipsoidal", or spherical cap - and then to correlate overall shape factors (e.g. height-to-width ratio for the ellipsoidal regime and wake angle for spherical-cap regime). It is worth pointing out that separate approaches have been found to be needed for predicting the shapes of liquid drops in gases than are used for drops and bubbles in liquids. the various methods are well covered elsewhere (Grace et al, 1976; Clift et al, 1978; Grace and Wairegi, 1986) and therefore will not be described further here.

CONCLUSION

The factors influencing the shapes of bubbles and drops are described and applied to the prediction of shapes. For stationary fluid particles, interfacial or surface tension forces act to retain a spherical shape, while hydrostatic forces act to cause deformation (flattening for a sessile fluid particle and stretching for a pendant one). for drops and bubbles in motion, dynamic pressure forces tend to cause flattening over the front part of the surface. A simple theoretical model is presented which provides good qualitative predictions for fluid particles at high Reynolds numbers. Further refinement of the approach is needed to improve the quantitative predictive powers of the model.

NOMENCLATURE

a	radius of sphere or semi-major axis of oblate spheroid
a_0	undeformed sphere radius
b_1, b_2	anterior, posterior semi-minor axis of oblate spheroid
d_e	volume-equivalent sphere diameter
Eo	Eotvos number, $g\Delta\rho d_e^2/\sigma_e$
Eo^*	Eotvos number based on 2a, $4g\Delta\rho a^2/\sigma$
e_1	eccentricity of anterior semi-spheroid
g	acceleration of gravity
K_3	function of e_1 defined by equation (23)
M	fluid property group defined by equation (15)
P_2, P_3	Legendre polynomials
p, p'	pressure on outside, inside of fluid particle
P_{dyn}	dynamic pressure, equation (16)
q	magnitude of local fluid velocity
R_0	radius of curvature at nose of fluid particle
R_1, R_2	principal radii of curvature at point on surface
Re, Re'	Reynolds number based on external, internal fluid properties as defined in equations (7) and (8)
r	radial coordinate

v_T	terminal settling or rising velocity
We	Weber number, $2a_0\rho v_T^2/\sigma$
We*	Weber number, $2a\rho v_T^2/\sigma$
z	vertical coordinate, measured away from nose level as shown in Figure 1.
$\Delta\rho$	magnitude of density difference, i.e. $\lvert \rho - \rho' \rvert$
η_1	elliptical coordinate angle
θ	polar coordinate
κ	viscosity ration, μ' / μ
λ	function of κ and ρ'/ρ defined by equation (14)
μ, μ'	viscosity of external, internal fluid
ρ, ρ'	density of external, internal fluid
σ	surface or interfacial tension
ψ, ψ'	external, internal Stokes stream function

REFERENCES

Akiyama, T., and Yamaguchi, K., Flow past a Spherical Drop or Bubble at Low Reynolds Number, Can. J. Chem. Eng., vol. 53, pp. 695-698, 1975.

Bashforth, F., and Adams, H., An Attempt to Test the Theories of Capillary Action, Cambridge, Univ. Press, London, 1883.

Best, A.C., The Shape of Raindrops and Mode of Disintegration of Large Drops, Met. Res. Committee Rept. MRP330, 1947.

Bhaga, D., and Weber, M.E., Bubbles in Viscous Liquids: Shapes, Wakes and Velocities, J. Fluid Mech., vol 105, pp. 61-85, 1981.

Brignell, A.S., The Deformation of a Liquid Drop at Small Reynolds Number, Quart. J. Mech. and App. Math., vol. 26, pp. 99-107, 1973.

Clift, R., Grace, J.R., and Weber, M.E., Bubbles, Drops and Particles, Academic Press, New York, 1978.

Davis, R.M. and Taylor, G.I., The Mechanics of Large Bubbles Rising through Extended Liquids and through Liquids in Tubes, Proc. Roy. Soc., vol. A200, pp. 375-390, 1950.

Davies, R.E., and Acrivos, A., The Influence of Surfactants on the Creeping Motion of Bubbles, Chem. Eng. Sci., vol. 21, pp. 681-685, 1966.

Finlay, B.A., A Study of Liquid Drops in an Air Stream, Ph.D. dissertation, Univ. of Birmingham, 1957.

Garner, F.H., and Skelland, A.H.P., Some Factors Affecting Droplet Behaviour in Liquid-Liquid Systems, Chem. Eng. Sci., vol. 4, pp. 149-158, 1955.

Grace, J.R., Shapes and Velocities of Bubbles Rising in Infinite Liquids, Trans. Instn. Chem. Engrs., vol. 51, pp. 116-120, 1973.

Grace, J.R., Wairegi, T., and Nguyen, T.H., Shapes and Velocities of Single Bubbles and Drops Moving Freely through Immiscible Liquids, Trans. Instn. Chem. Engrs., vol. 54, pp. 255-264, 1976.

Grace, J.R., and Wairegi, T., Properties and Characteristics of Drops and Bubbles, in Encyclopedia of Fluid Mechanics, ed. N.P. Cheremisinoff, Gulf Publishing Co., vol. 4, pp. 43-57, 1986.

Green, A.W., An approximation for the Shape of Large Raindrops, J. Appl. Meteorology, vol. 14, pp. 1578-1583, 1975.

Haberman, W.L., and Morton, R.K., An experimental Study of Bubbles Moving in Liquids, David Taylor Model Basin Rept. No. 802, 1953.

Hadamard, J.S., Mouvement Permanent Lent d'une Sphère Liquide et Visqueuse dans un Liquide Visqueux, Compt. Rend. Acad. Sci., vol. 152, pp. 1735-1738, 1911.

Hartland, S., and Hartley, R.W., Axisymmetric Fluid-Liquid Interfaces, Elsevier, Amsterdam, 1976.

Huang, S.W., Ph.D. dissertation, Illinois Inst. of Technology, Chicago, 1968.

McDonald, J.E., The Shape and Aerodynamics of Large Raindrops, J. Meteor., vol. 11, pp. 478-494, 1954.

Pan, F.Y., and Acrivos, A., Shape of a Drop or Bubble at Low Reynolds Number, Ind. Eng. Chem. Fund., vol. 7, pp. 227-232, 1968.

Pruppacher, H.R., and Beard, K.G., A Wind Tunnel Investigation of the Internal Circulation and Shape of Water Drops Falling at Terminal Velocity in Air, Quart. J. Roy. Meteor. Soc., vol 96, pp. 247-256, 1970.

Pruppacher, H.R. and Pitter, R.L., A Semi-Empirical Determination of the Shape of Cloud and Rain Drops, J. Atmo. Sci., vol. 28, pp. 86-94, 1971.

Rybczynski, E., Über die fortschreitende Bewegung einer flüssigen Kugel in einem zähen Medium, Bull. Int. Acad. Sci. Cracovic (Ser. A.), pp. 40-46, 1911.

Saffman, P.G., On the Rise of Small Air Bubbles in Water, J. Fluid Mech., vol. 1, pp. 249-275, 1956.

Savic, P., Circuation and Distortion of Liquid Drops Falling through a Viscous Medium, Nat. Res. Council Canada Rept. NRC-MT-22, 1953.

Taylor, T.D., and Acrivos, A., On the Deformation and Drag of a Falling Viscous Drop at Low Reynolds Number, J. Fluid Mech., vol. 18, pp. 466-476, 1964.

Wairegi, T., The Mechanics of Large Drops and Bubbles Moving through Extended Liquid Media, Ph.D. dissertation, McGill Univ., Montreal, 1974.

Wairegi, T., and Grace, J.R., The Behaviour of Large Drops in Immiscible Liquids, Intern. J. Multiphase Flow, vol. 3, pp. 67-77, 1976.

Wellek, R.M., Agrawal, A.K., and Skelland, A.H.P., Shape of Liquid Drops Moving in Liquid Media, A.I.Ch.E.J., vol. 12, pp. 854-862, 1966.

Chapter 7

Flow of Particles in Viscoelastic Fluids

P. N. KALONI and J. STASTNA
Department of Mathematics and Statistics
University of Windsor
Windsor, Ontario, N9B 3P4 Canada

INTRODUCTION

In this chapter we are concerned with two special problems in the flow of particles in viscoelastic fluids: the flow of viscoelastic fluids past fluid spheres and the calculation of effective viscosity of a suspension in a non-Newtonian fluid.

The flow of liquids past submerged objects, e.g., particles, drops and bubbles, at low Reynolds number is one of the oldest classical problems of fluid mechanics. The first important analysis of this problem dates to 1851 when Stokes [1] investigated the translation of a rigid sphere through an unbounded quiescent fluid at zero Reynolds number. A review of the researches published prior to 1965, which deal mainly with a Newtonian fluid, can be found in the book by Happel and Brenner [2]. Subsequent developments, both in Newtonian and non-Newtonian fluids have appeared in the book by Clift, Grace and Weber [3] and in various review articles by Goldsmith and Mason [4], Leal [5], Caswell [6], Brunn [7], Dairanieh and McHugh [8] and Chhabra [9].

Another long-standing problem in fluid mechanics has been the calculation of the effective viscosity of a suspension of small particles. Einstein [10] considered the dilute suspension of solid spherical particles in a viscous fluid and obtained an expression for the effective viscosity. Einstein's work indicates the suspension to be Newtonian with increased viscosity. In recent years it has become increasingly clear that the fluid properties of a suspension are inherently non-Newtonian and that the knowledge of an effective viscosity alone can not explain the complete suspension behaviour. Review articles by Batchelor [10a] Brenner [11], Jinescu [12] and Jeffery and Acrivos [13] discuss some important features concerning the flow properties of suspensions.

The slow flow of viscoelastic liquids past submerged objects is an idealization of flows used in various industrial processes. As an example one can mention the use of slender or rod shape objects as seeding sites during the processing of polymer solutions, melts and blends [8]. In many flows a better understanding of the phenomenon of the deformation of droplets suspended in a liquid is necessary. The bulk rheological properties of such systems are certainly affected by the deformation of droplets as has been demonstrated in [14, 15,16]. These are just examples of more general theoretical problems - the understanding of the relationship between the material properties of liquids in viscometric flows and the behavior of these liquids in various non-viscometric, industrial flows. We have no assurance that the constitutive equations which are useful in viscometric flows can describe properly the same materials in more complex flows. Of various problems associated with the above mentioned flows, the translation of a spherical particle in a non-Newtonian fluid, is one of the simplest, interesting problem.

The Classical Hadamard-Rybczynski Problem and its Generalization

Assume that we study the translation of a particle in a fluid. The flow is assumed to be steady, incompressible, axisymetric and at low Reynolds number. Using the bar for the interior (particle) phase variables we can write the dimensionless constitutive equations:

$$\bar{T} = \bar{P}I + 2\kappa\bar{D} , \quad (\text{viscosity } \mu) \quad - \text{particle} \tag{1}$$

$$T = -PI + 2D , \quad (\text{viscosity } \eta_o) \quad - \text{surrounding medium} \tag{2}$$

Here $\bar{T}(T)$ represents the stress tensor, $\bar{D}(D)$ is the rate of deformation tensor, $\bar{P}(P)$ is the pressure, I the unit tensor and κ is the ratio of particle viscosity μ to η_o, the viscosity of exterior fluid. The appropriate field equations are:

$$-\nabla\bar{P} + \kappa\Delta\bar{V} = \frac{\bar{\rho}}{\rho} R_e \bar{V} \cdot \nabla\bar{V} \tag{3}$$

$$-\nabla P + \Delta V = R_e V \cdot \nabla V, \tag{4}$$

where V in the dimensionless gradient, $\bar{v}(v)$ represents the velocity $\bar{\rho}(\rho)$ is the density and R_e is the Reynolds number defined as

$$R_e = \rho \frac{Ua}{\eta_o} , \tag{5}$$

with a the radius of a spherical particle and U - the Hadamard-Rybczynski terminal velocity.

In Stokes approximation ($R_e \to o$) the system (3), (4) can be solved with the help of stream functions $\bar{\psi}(\psi)$ and the classical boundary conditions [2,3]. The same scheme can also be used in case of more complicated constitutive equations. The problem of the translation of a single particle in a viscoelastic fluid is usually studied in the flow regimes with the time scale much larger than the characteristic relaxation time of the viscoelastic fluid. In these flows the Weissenberg number W_e, which is the ratio of a characteristic relaxation time λ for the fluid and the convective time scale a/U, is very small. Then the perturbation method is a power series expansion in terms of the Weissenberg number $W_e = U\lambda/a$, can be used, if the following conditions hold

$$\rho \frac{Ua}{\eta_o} \equiv R_e << \frac{U\lambda}{a} \equiv W_e < 1. \tag{6}$$

In this limit, general constitutive equations reduce to the n-th order fluid models which are usually used in the equations of motion. These models are also recovered from regular perturbations, in the low W_e limit, of Oldroyd differential models. Leslie [17] solved the problem of the translation of a solid spherical particle in a viscoelastic fluid modelled as a 6 constant Oldroyd fluid [18]. Leslie calculated the terminal velocity to the second order in W_e, and found (due to the shear thinning of the used Oldroyd model) that the drag on the solid sphere in a viscoelastic fluid is always less than the corresponding Stokes drag. Similar result, for a third order fluid, has been obtained by Giesekus [19]. Tiefenbruck and Leal [20] calculated the terminal velocity of a spherical bubble rising in a viscoelastic fluid modelled as a 4-constant Oldroyd fluid with a variable time derivative. In contrast to the result for a solid sphere, the terminal velocity of a gas bubble can be less or greater than the corresponding Hadamard-Rybczynski bubble velocity, depending upon the "type" of the variable time derivative. These results are in agreement with the findings of Ajayi [21] and Shirotsuka and Kawase [22].

To demonstrate the power series expansion let us consider the the translation of a Newtonian (originally a spherical) droplet through a viscoelastic fluid described by an Oldroyd-type constitutive equation (dimensionless form):

$$\mathbf{T} = -P\mathbf{I} + \tau \tag{7}$$

$$\tau + W_e \{ \mathbf{V} \cdot \nabla \tau + \mathbf{W} \cdot \tau - \tau \cdot \mathbf{W} - \alpha (\mathbf{D} \cdot \tau + \tau \cdot \mathbf{D}) \} =$$

$$2\mathbf{D} + 2\varepsilon W_e \{ \mathbf{V} \cdot \nabla \mathbf{D} + \mathbf{W} \cdot \mathbf{D} - \mathbf{D} \cdot \mathbf{W} - 2\alpha \mathbf{D} \cdot \mathbf{D} \} , \tag{8}$$

$$2\mathbf{D} = (\nabla \mathbf{V} + (\nabla \mathbf{V})^T), \quad 2\mathbf{W} = (\nabla \mathbf{V} - (\nabla \mathbf{V})^T) \tag{9}$$

where $W_e = \frac{\lambda_1 U}{a}$, $\varepsilon = \frac{\lambda_2}{\lambda_1}$, and λ_1, λ_2 are relaxation and retardation times, respectively. The paramater α lies in the interval $[0,1]$. The qualitative rheological behavior of real viscoelastic fluids is somewhere between the value $\alpha=0$ and $\alpha=1$. The models with intermediate time derivatives belong, for instance, the models of Phan-Thien and Tanner [23] and Johnson and Segalman [24]. These are network-type models with non-affine deformation, proposed for polymer melts. Similar models have been proposed for dilute polymer solutions and suspensions (25,26). The constitutive equation for a Newtonian droplet has the form given by equation (1) where κ is the ratio of droplet viscosity μ to η_o, the zero shear rate of viscosity of the exterior fluid.

The equations of motion are:

particle: $\quad - \nabla \bar{P} + \kappa \Delta \bar{\mathbf{V}} = \frac{\bar{\rho}}{\rho} R_e \bar{\mathbf{V}} \cdot \nabla \bar{\mathbf{V}} ,$ \hfill (10)

surrounding medium: $-\nabla P + \mathbf{V} \cdot \tau = R_e \mathbf{V} \cdot \nabla \mathbf{V}.$ \hfill (11) where τ is given by the constitutive equation (8).

The time scale of the considered flow is given as the time needed by a fluid particle to orbit the droplet. This time is proportional to a/U, and we assume that the fluid element stress relaxes much faster than a a/U, i.e., $\lambda_1 < a/U$, or $W_e < 1$. In this limit one assumes that all kinematic and stress variables can be expanded in W_e. Then, for example,

$$\mathbf{V} = \sum_{n=o}^{\infty} \mathbf{v}^{(n)} (W_e)^n \implies \tau = \sum_{n=o}^{\infty} \tau^{(n)} (W_e)^n \;,$$

which yields

$$\tau^{(n)} = 2\mathbf{D}^{(n)} + \sum_{k=o}^{n-1} \{ \mathbf{v}^{(k)} \cdot \nabla (2\varepsilon \mathbf{D}^{(n-k-1)} - \tau^{(n-k-1)})$$

$$+ (\mathbf{W}^{(k)} - \alpha \mathbf{D}^{(k)}) \cdot (2\varepsilon \mathbf{D}^{(n-k-1)} - \tau^{(n-k-1)})$$

$$- (2\varepsilon \mathbf{D}^{(k)} - \tau^{(k)}) \cdot (\alpha \mathbf{D}^{(n-k-1)} + \mathbf{W}^{(n-k-1)}) \};$$

$$n = 1,2 \dots \quad . \tag{12}$$

We can see that for the considered Oldroyd model it is possible to decompose the extra stress tensor into a Newtonian and non-Newtonian parts

$$\tau = 2\mathbf{D} + \mathbf{N} \tag{13}$$

The mentioned asymptotic expansion (in powers of W_e) for instance yields, the relatively simple, first approximation

$$\tau^{(1)} = 2\mathbf{D}^{(1)} + 2(\varepsilon-1)[\mathbf{v}^{(o)} \nabla \cdot \mathbf{D}^{(o)} + \mathbf{W}^{(o)} \cdot \mathbf{D}^{(o)} - \mathbf{D}^{(o)} \cdot \mathbf{W}^{(o)}$$

$$- 2\alpha \mathbf{D}^{(o)} \cdot \mathbf{D}^{(o)}] \tag{14}$$

The same asymptotic expansion in equations (10) and (11) yields an iterative system of equations of motion. Since even if there is no flow the stress tensors are not zero but reduce to hydrostatic isotropic terms one has the following expansions for dimensionless $\bar{P}$ and P

$$\bar{P} = \sum_{n=-1}^{\infty} \bar{P}^{(n)} (W_e)^n, \quad P = \sum_{n=-1}^{\infty} P^{(n)} (W_e)^n. \tag{15}$$

Writing

$$\frac{\bar{\rho}}{\rho} R_e = \frac{\bar{\rho} a^2}{\eta_o \lambda_1} W_e = \bar{A} W_e \tag{16}$$

and

$$R_e = \frac{\rho a^2}{\eta_o \lambda_1} W_e = A W_e \tag{17}$$

we obtain for Newtonian particle

149

$$-\boldsymbol{\nabla}\bar{P}^{(-1)} = 0$$

$$-\boldsymbol{\nabla}\bar{P}^{(o)} + \kappa\Delta\bar{\mathbf{V}}^{(o)} = 0 \tag{18}$$

$$-\boldsymbol{\nabla}\bar{P}^{(n)} + \kappa\Delta\bar{\mathbf{V}}^{(n)} = \bar{A}\sum_{k=0}^{n-1} \bar{\mathbf{V}}^{(k)}\cdot\boldsymbol{\nabla}\bar{\mathbf{V}}^{(n-k-1)}, \text{ for } n=1,2,3$$

and for surrounding Oldroyd medium

$$-\nabla P^{(-1)} = 0$$

$$-\nabla P^{(o)} + \boldsymbol{\nabla}\cdot\tau^{(o)} = 0 \qquad -\nabla P^{(o)} + \Delta\mathbf{V}^{(o)} = 0$$

$$-\nabla P^{(n)} + \boldsymbol{\nabla}\cdot\tau^{(n)} = A\sum_{k=0}^{n-1} \mathbf{V}^{(k)}\cdot\boldsymbol{\nabla}\mathbf{V}^{(n-k-1)}, \text{ for } n=1,2,3, \tag{19}$$

where $\tau^{(n)}$ is given by (12).

This iteration scheme has to be accompanied by the appropriate expansion of boundary conditions. One has to note that usually it is assumed that the shape of the particle remains spherical at least to a first approximation. The change in shape has been considered for example in [16] and [17].

A slightly different general method of solution has been used in [28]. The starting point of this method is the decomposition (13). Since the flow is incompressible and axisymetric, one can introduce streamfunctions $\bar{\psi}$ and ψ, which satisfy identically the continuity equation in spherical coordinate system (origin at the particle center). Then

$$\bar{v}_r = -\frac{1}{r^2\sin\theta}\frac{\partial\bar{\psi}}{\partial\theta}, \quad \bar{v}_\theta = \frac{1}{r\sin\theta}\frac{\partial\bar{\psi}}{\partial r}. \tag{20}$$

Neglecting the inertial terms one has the known equation of motion for a particle phase

$$E^4\bar{\psi} = 0 \tag{21}$$

where

$$E^2 \equiv \frac{\partial^2}{\partial r^2} + \frac{\sin^2\theta}{r^2}\frac{\partial}{\partial\theta}\left(\frac{1}{\sin\theta}\frac{\partial}{\partial\theta}\right). \tag{22}$$

The solution of this homogeneous problem is well known [2].

Using the decomposition (13) one can write the following equation of motion for the surrounding Oldroyd fluid

$$E^4\psi = M(r,\theta) \tag{23}$$

where

$$M(r,\theta) = \left[\frac{\partial}{\partial\theta}(\mathbf{V}\cdot\mathbf{N})_r - \frac{\partial}{\partial r}(r(\mathbf{V}\cdot\mathbf{N})_\theta\right]\sin\theta,$$

Here the subscripts r and θ represent the r,θ components of the vector (**V.N**). The boundary conditions used in [28] are: the continuity of tangential velocity of the interface.

$$\frac{\partial \psi}{\partial r} = \frac{\partial \bar{\psi}}{\partial r} \quad \text{at } r = 1; \tag{24}$$

the continuity of tangential stress at the particle surface

$$\tau_{r\theta} = \bar{\tau}_{r\theta} \quad \text{at } r = 1; \tag{25}$$

the zero normal velocity at the particle surface (constant shape)

$$\psi = \bar{\psi} = 0 \quad \text{at } r = 1; \tag{26}$$

finite velocities at r=0, and uniform free velocity far from the particle

$$\lim_{r\to\infty} \frac{\bar{\psi}}{r^2} \text{ finite} \tag{27}$$

$$\psi = \frac{1}{2} \frac{\bar{U}}{U} \sin^2\theta \quad \text{as } r\to\infty \, , \tag{28}$$

where $\bar{U}$ is the terminal velocity.

Using the asymptotic expansions

$$\psi = \psi^{(o)}(r,\theta) + W_e \psi^{(1)}(r,\theta) + W_e^2 \psi^{(2)}(r,\theta) + \dots$$

$$\tau = \tau^{o}(r,\theta) + W_e \tau^{(1)}(r,\theta) + W_e^2 \tau^{(2)}(r,\theta) + \dots \tag{29}$$

$$\bar{U} = U + W_e u^{(1)}(r,\theta) + W_e^2 u^{(2)}(r,\theta) + \dots$$

and the decomposition (13)

$$\tau^{(k)} = 2D^{(k)} + N^{(k)},$$

one has to solve the equations

$$E^4 \psi^{(k)} = M^{(k)}(\psi^{(k-1)} \dots \psi^{(o)})$$

$$M^{(k)}(r,\theta) = [\frac{\partial}{\partial\theta}(V\cdot N^{(k)})_r - \frac{\partial}{\partial r}(r(V\cdot N^{(k)})_\theta)]\sin\theta \, . \tag{30}$$

accompanied by the interated boundary condition (24)-(28). The general solution of the problem (30) is given in [28].

For undeformed particle the normal stress balance can be written in the form [28]:

$$F_z^{(k)} = 2\pi \int_o^\pi [(-P^{(k)} + \tau_{rr}^{(k)})_{r=1} \cos\theta - \tau_{r\theta}^{(k)}|_{r=1} \sin\theta]\sin\theta d\theta. \tag{31}$$

In analogy with Newtonian case [2] one can express the pressure gradiant of the viscoelastic fluid as a function of the vorticity and the non-Newtonian stress

$$\nabla P^{(k)} = -\nabla \times \xi^{(k)} + \nabla . \mathbf{N}^{(k)} \tag{32}$$

where for axisymetric flow

$$\xi^{(k)} = E^2 \psi^{(k)} / r \sin^2 \theta . \tag{33}$$

Integrating the pressure integral (31) Quintana, et al., [28] have obtained the force balance equation from which the k-th order viscoelastic correction to the Hadamard-Rybczynski terminal velocity can be obtained. It has been found in [28] that the first viscoelastic correction to the Hadamard-Rybczynski terminal velocity is zero, i.e., the higher order corrections are necessary. These authors report the non-zero second correction $\bar{U}^{(2)}/U$, which can be negative or positive depending upon the values of parameters α and κ. The limiting cases of a solid sphere ([17]) and a spherical gas bubble [20,27] are in agreement with the results of these authors.

The outlined procedure is sufficiently general and can be modified to obtain the perturbations for various possible cases, viscoelastic particles, spheres, bubbles, etc. In the case of deformable particles one has to be more careful in formulating the boundary conditions.

The boundary conditions on the interface of a deformable particle can be summarized as follows:

Assuming that particle is only slightly deformable, the surface of the particle can be described by the following dimensionless equation (spherical coordinates):

$$r = 1+\beta(\theta), \quad |\beta(\theta)| \ll 1, \quad |\frac{d\beta}{d\theta}| \ll 1, \tag{34}$$

and the unit normal vector can be approximated by the vector

$$\mathbf{n} = (1, -\frac{1}{r}\frac{d\beta}{d\theta}, 0). \tag{35}$$

Then on $r = 1+\beta(\theta)$ one has to satisfy

1. $\bar{\mathbf{V}}.\mathbf{n} = \mathbf{V}.\mathbf{n} = 0$

2. $\bar{\mathbf{V}} - (\bar{\mathbf{V}}.\mathbf{n})\mathbf{n} = \mathbf{V} - (\mathbf{V}.\mathbf{n})\mathbf{n}.$

3. $\bar{T}.\mathbf{n} = -[(\bar{T}.\mathbf{n}).\mathbf{n}]\mathbf{n} = \mathbf{T}.\mathbf{n} - [(\mathbf{T}.\mathbf{n}).\mathbf{n}]\mathbf{N},$ $\qquad$ (36)

4. $(\mathbf{T}.\mathbf{n}).\mathbf{n} - (\bar{T}.\mathbf{n}).\mathbf{n} = \frac{\sigma'}{\eta_o U}(\frac{1}{R_1} + \frac{1}{R_2}).$

Here R_1 and R_2 are the principal radii of curvature of the particle, and σ' is the surface tension. Introducing $\sigma \equiv \frac{\sigma^1 \lambda_1}{\eta_o a}$ and using the relation [29]

$$\frac{1}{R_1} + \frac{1}{R_2} = 2 - 2\beta - \cot\theta \frac{d\beta}{d\theta} - \frac{d^2\beta}{d\theta^2} \, , \tag{37}$$

one can obtain the following conditions on the surface $r = 1 + \beta(\theta)$:

$$v_r = \frac{v_\theta}{r} \frac{d\beta}{d\theta} \, , \quad \bar{v}_r = v_r \, , \quad \bar{v}_\theta = v_\theta \, , \quad \bar{\tau}_{r\theta} = \tau_{r\theta} \, , \tag{38}$$

$$\tau_{rr} - \tau_{\theta\theta} = \bar{\tau}_{rr} - \bar{\tau}_{\theta\theta}$$

and

$$\bar{P} - P + \tau_{rr} - \bar{\tau}_{rr} = \frac{\sigma}{W_e}(2 - 2\beta - \cot\theta \frac{d\beta}{d\theta} - \frac{d^2\beta}{d\theta^2}) . \tag{39}$$

In addition the condition of constant volume yields

$$\int_{-1}^{1} \beta(z)dz = 0, \quad z = \cos\theta, \tag{40}$$

and if one postulates the origin of the coordinate system at the centroid of the particle, another condition appears:

$$\int_{-1}^{1} z\beta(z)dz = 0. \tag{41}$$

It is clear, now, that one has to apply the perturbation scheme (expansion in powers of W_e) also in the condition (39). The situation can be very complicated if the surface tension is not a constant.

Theoretical predictions of a small downstream shift, in the slow uniform flow of viscoelastic liquids past submerged objects [17,19,27,28], and a decrease in the drag at $0(W_e^2)$, are difficult to verify experimentally. It has been pointed out in [30] that the effects described as viscoelastic can be also explained as non-Newtonian viscosity effects. Numerical simulation, the drag force measurement and the flow visualization described in [30] lead to the hypothesis that the non-Newtonian viscosity can be responsible for the friction and pressure drag decrease and larger separating of vortices behind the sphere. No shift in the streamlines has been attributed to the non-Newtonian viscosity. On the other hand the same authors believe that due to the presence of viscoelasticity, the normal force drag increases and the separating vortices behind the sphere become smaller, and there is an upstream shift in the streamlines. As far as viscoelastic effects are concerned similar observations have been made in [20] and some experimental evidence of a small upstream shift has been given in [31]. The problem of creeping flow of various non-Newtonian fluids past a sphere has been extensively studied prior to 1980, see e.g., [32,33,34]. Tanner and coworkers [35,36] have studied the drag on a sphere in a power law fluid and in a fluid modelled by the modified Phan-Thien-Tanner fluid. From their numerical simulations they have concluded that even through for most

polymeric liquids it is difficult to distinguish the pure
viscous shear-thinning and pure elastic effect, there is a
general belief that the viscosity effects are usually more
important. In [36], using again the numerical simulation,
it is been found that the drag coefficient is remarkably
reduced by the fluid elasticity and the shear-thinning viscosity. Also a shift of the streamlines in the downstream
direction is reported there. Mena, Manero and Leal]37],
however, on the basis of their experiment with four different
type of fluids, conclude that elastic effects are of primary
importance at extremely low shear rates and that, for high
shear rates, shear thinning effects become predominant.
Experimental evidence for a drastic drag reduction in non-
shear-thinning elastic liquids has been reported by Chhabra
et al. [38]. Chilcott and Rallison [39] modelled a dilute
polymer solution as a suspension of dumbbells with finite
extensibility. Using no-slip, and a zero-tangential-stress
boundary conditions at the body surface, they calculated the
time-dependent flow past cylindrical and spherical surfaces
at low Reynolds number. Numerically stable results have been
obtained up to the value 16 of Deborah number. The reported
drag reduction of 25% by Chabbra et al. is much larger than
the drop of order 5% obtained in the work [39]. Recent
studies of Chmielewski et al. [40], which deal with the
experimental measurements in non-shear-thinning elastic
liquids, do not support the findings of [38], but do support,
in part, of [39].

In a series of papers Barthes-Biesel and coworkers have
studied the motion and deformation of a spherical capsule,
filled with an incompressible Newtonian fluid, (steady and time
dependent) [41,42,43]. A regular perturbation solution of the
general problem with a spherical particle which undergoes
small deformations has been obtained in [43]. It has been
shown there, that with a purely viscous membrane the capsule
deforms into an ellipsoid and has a continuous flipping
motion. When the membrane relaxation time is of the same
order as the shear time, the particle reaches a steady ellipsoidal shape which is oriented with respect to streamlines
at an angle between 45° and 0°; the angle decreases with
increasing shear rate. It has been predicted that the deformation can reach a maximum value. When the particle
viscosity is much higher than that of the suspending medium,
an oscillatory motion is predicted. The relevance of the
model to the behavior of red blood cells is also discussed
in [43].

We close this section by providing an elegant formula, for
determining the low Reynolds number drag $\mathbf{F}$ on an obstacle
(which may be a rigid, or stress free or a liquid droplet)
immersed in a non-Newtonian fluid [39]. Let us suppose that
the constitutive equation of the non-Newtonian fluid can be
written as

$$\sigma = -p\mathbf{I} + 2\mu_s \mathbf{D} + \sigma^{NN} \tag{42}$$

where σ^{NN} is the non-Newtonian contribution. Then

$$F = F_s + \int_{\hat{V}} D_s : \sigma^{NN} dV \ , \tag{43}$$

where F_s is the solvent drag force and D_s is the Stokes rate of strain tensor and $\hat{V}$ is the volume occupied by the non-Newtonian fluid. In this manner one can avoid calculating pressure.

STEADY RHEOLOGICAL BEHAVIOUR OF DILUTE SUSPENSIONS
IN VISECOELASTIC FLUIDS

In this section we now consider the phenomena associated with shearing motion of the fluid relative to suspended particle. In general, the problems in suspension can either be handled via mixture theories or by considering the suspension as homogeneous fluids and ascribing to them certain effective fluid properties. This latter situation is applicable when the particle dimensions are much smaller compared with the dimensions of the apparatus containing the suspension. In defining effective viscosity of the suspension, as Happel and Brenner [2] point out a number of variables are involved: (i) the nature of the fluid; (ii) the nature of the suspended particles; (iii) the concentration of suspended particles, and (iv) the motion of particles and the fluid. Also a number of methods have been applied to determine the effective viscosity of suspensions (mostly for dilute cases) see [2,11,14] and here we discuss one such method only.

Landau and Lifshitz [29], in their book, proposed a new derivation of the Einstein's viscosity formula for the dilute suspension of rigid spherical particles in a Newtonian fluid.

In this work, the authors employed volume averaging of the stress tensor and showed that such an average can be found without the knowledge of the stress inside the particles. By relating the averaged stress tensor to the averaged rate of strain tensor, a constitutive equation was derived and the expression for the effective viscosity was calculated. Batchelor [44], inspired with the ideas of [29], refined and improved the method of Landau and Lifshitz [29] and applied it not only to non-spherical particles [44] but to a variety of related problems [45]. In the following we follow Batchelor's approach to treat the dilute suspension of rigid spherical particles in a viscoelastic fluid. Works on Brownian particle suspensions in Newtonian and non-Newtonian fluids have been summarized by Brenner [46] and Hill and Soane [47], respectively.

Consider a dilute suspension of rigid spherical particles of identical size and shape dispersed in an incompressible non-Newtonian fluid. Let V be the volume of the suspension type enough to contain many particles but small enough with respect to the length scale of the flow. Let V_O and A_O be the volume and surface area of a spherical particle which is in V. We define the stress average $\langle \sigma_{ij} \rangle$ by [29]

$$\langle \sigma_{ij} \rangle \equiv \frac{1}{V} \int \sigma_{ij} dV \tag{44}$$

where in defining (44) it is assumed that inertia effects in the suspension are negligible. Following Batchelor [44] we also define

$$\frac{\partial U_i}{\partial x_j} = \frac{1}{V} \int \frac{\partial u_i}{\partial x_j} dV = \langle \frac{\partial u_i}{\partial x_j} \rangle$$

$$\langle D_{ij} \rangle = \frac{1}{V} \int D_{ij} dV$$

(45)

$$\langle D_{ik} D_{kj} \rangle = \langle D_{ik} \rangle \cdot \langle D_{xj} \rangle$$

where in writing the last equation (45) we have assumed that the fluctuating part of $(D_{ij})^2$ should be quite small as it should be of order of the square of the volume fraction of the rigid spheres in the suspension. Further we shall also use the identities of the type

$$\int \frac{\partial v_i}{\partial x_j} dV = \int v_i n_j dA, \quad \sigma_{ij} = \frac{\partial(\sigma_{ik} x_k)}{\partial x_j} ,$$

(46)

$$\int_{V_o} \sigma_{ij} dV = \int_{A_o} \sigma_{ik} x_j n_k dA - \int_{V_o} \frac{\partial \sigma_{ik}}{\partial x_j} dV .$$

We assume that the ambient fluid is a non-Newtonian fluid modelled by a second order fluid, in which case

$$\sigma_{ij} = -p\delta_{ij} + \mu_o A_{ij}^{(1)} + \alpha_1 A_{ik}^{(1)} A_{kj}^{(1)} + \alpha_2 A_{ij}^{(2)}$$

(47)

where

$$A_{ij}^{(1)} = D_{ij} = (\frac{\partial v_i}{\partial x_j} + \frac{\partial v_j}{\partial x_i})$$

$$A_{ij}^{(2)} = A_{ij}^{(1)} + A_{ik}^{(1)} v_{k,j} + A_{jk}^{(1)} v_{k,i} .$$

In (47) μ_o, α_1 and α_2 are constants. If we non-dimensionalize equation (47) by a characteristic length a and a characteristic velocity Ga where G is the average shear rate of suspension, we can write (47) as

$$\sigma_{ij} = - P\sigma_{ij} + A_{ij}^{(1)} + \lambda[A_{ik}^{(1)} A_{kj}^{(1)} + \varepsilon A_{ij}^{(2)}]$$

(48)

where

$$\lambda = \frac{\alpha_1 G}{\mu_o} \quad \text{and} \quad \varepsilon = \frac{\alpha_2}{\alpha_1} .$$

On applying the definition (44) to (48) we can write

$$\langle\sigma_{ij}\rangle = \frac{1}{V}\int_{V-\Sigma V_o} [-P\delta_{ij} + A_{ij}^{(1)} + \lambda\{A_{ik}^{(1)}A_{kj}^{(1)} + \varepsilon A_{ij}^{(2)}\}]dV$$

$$+ \frac{1}{V}\Sigma\int_{V_o} [-P\delta_{ij} + A_{ij}^{(1)} + \lambda\{A_{ik}^{(1)}A_{kj}^{(1)} + \varepsilon A_{ij}^{(2)}\}]dV \qquad (49)$$

where $V-\Sigma V_o$ is the volume occupied by the fluid. Further application of (45)-(46) to (49) leads to [48]

$$\langle\sigma_{ij}\rangle = -\bar{P}\delta_{ij} + \langle A_{ij}^{(1)}\rangle + \lambda[\langle A_{ik}^{(1)}A_{kj}^{(1)}\rangle + \varepsilon A_{ij}^{(2)}]$$

$$- \frac{1}{V}\Sigma\int_{V_o}\lambda(A_{ik}^{(1)}A_{kj}^{(1)} + \varepsilon A_{ij}^{(2)})dV + \frac{1}{V}\Sigma\int_{A_o}[\sigma_{ik}n_k x_j - \mu_o(v_i n_j)$$

$$+ v_j n_i)]dA . \qquad (50)$$

We note that for $\lambda=0$ the above reduces to the expression given in [44]

With the origin of the coordinate at the centre of the drop, we now assume that all variables can be expanded in powers of λ, i.e.,

$$v_i = v_i^{(0)} + \lambda v_i^{(1)} + \lambda^2 v_i^{(2)} + \ldots$$

$$\langle\sigma_{ij}\rangle = \langle\sigma_{ij}^{(0)}\rangle + \lambda\langle\sigma_{ij}^{(1)}\rangle + \lambda^2\langle\sigma_{ij}^{(2)}\rangle + \ldots \qquad (51)$$

and so on. Here the zero[th] order term correspond to Newtonian problem which has been solved in [10,29,44]. It was pointed out in [44] that in order to find the effect of the last expression in (50), one could simply determine those values for a single sphere and then sum them for n spheres in a representative volume. Moreover in order to evaluate the values of those integrals on the surface of a sphere we need to solve the boundary value problem of the shear flow past a sphere in a second order fluid. This problem essentially has been solved by Peery [49]. Peery [49] considers the solution of the boundary value problem:

$$\frac{\partial^2 v_i}{\partial x_k \partial x_k} - \frac{\partial p}{\partial x_i} = -\lambda\frac{\partial}{\partial x_k}[A_{ip}^{(1)}A_{pk}^{(1)} + \varepsilon A_{ik}^{(2)}]$$

$$\frac{\partial v_i}{\partial x_i} = 0 \qquad (52)$$

$$v_i = 0 \text{ on } r=a$$

$$v_i = \alpha_{ij}x_j \text{ as } r \to \infty$$

$$p = 0$$

where α_{ij} is constant second order tensor with $\alpha_{ii}=0$. In short the group of equations to be solved are:

$$\frac{\partial^2 v_i^{(0)}}{\partial x_k \partial x_k} = \frac{\partial p^{(0)}}{\partial x_i} \quad , \quad \frac{\partial v_i^{(0)}}{\partial x_i} = 0 \quad ,$$

$$v_i^{(0)} = 0 \quad \text{on } r=1, \quad v_i^{(0)} = \alpha_{ij} x_j$$
$$\text{as } r \to \infty \quad ,$$
$$p^{(0)} = 0$$

(53)

and

$$\frac{\partial^2 v_i^{(1)}}{\partial x_k \partial x_k} - \frac{\partial p^{(1)}}{\partial x_i} = -\frac{\partial}{\partial x_k}[N_{ik}^{(0)}] , \quad \frac{\partial v_i^{(1)}}{\partial x_i} = 0$$

(54)

$$v_i^{(1)} = 0 \text{ as } r=1, \quad v_i^{(1)} = 0$$
$$\text{as } r \to \infty$$
$$p^{(1)} = 0$$

and so on, where $N_{ik} = [A_{ip}^{(1)} A_{pk}^{(1)} + A_{ik}^{(2)}]$.

Thus by solving the Newtonian problem (53) one finds $v_k^{(0)}$ and $p^{(0)}$. These values are then substituted in (54) and the solutions $v_k^{(1)}$ and $p^{(1)}$ are obtained. The process can be continued up to the desired order solution in λ.

Kaloni and Stastna [48] calculated the values for $\langle\sigma_{ij}\rangle$ only up to order λ and found that up to this order there was no effect in the effective (shears) viscosity. They did, however, found effect in the normal stress coefficients. A more general treatment when the suspended particles are Newtonian fluid drops, again suspended in a second order fluid, has been carried out by Sun and Jayaraman [50] and we report some of their results.

These authors proceed with similar assumption to those of [48] but simplify an expression in $A_{ij}^{(2)}$ by using the identity (using direct notation)

$$\mathbf{V} \cdot (\mathbf{VA}_1) = \mathbf{V} \cdot \mathbf{VA}_1 + (\nabla \cdot \mathbf{V})\mathbf{A}_1 .$$

Since $\mathbf{V} \cdot \mathbf{V}=0$, the volume integral of $\mathbf{V} \cdot (\mathbf{VA}_1)$ can then be written as a surface integral. Thus in place of (50), they arrive at

$$\langle\sigma\rangle = -\bar{P}\mathbf{I} + \langle\mathbf{A}_1\rangle + \lambda\langle\mathbf{A}_1 . \mathbf{A}_i\rangle + \lambda\epsilon\langle\mathbf{A}_2\rangle$$

$$- \frac{1}{V}\Sigma\int_{V_o} \{\lambda\epsilon(\mathbf{A}_1 . \nabla\mathbf{V} + \nabla\mathbf{V}^\Gamma\mathbf{A}_1) + \lambda\mathbf{A}_1 . \mathbf{A}_1\}dv$$

$$+ \frac{1}{V} \Sigma\int_{A_o} \{\mathbf{n} . (\sigma\mathbf{x} - \lambda\epsilon\mathbf{VA}_1) - (\mathbf{n}\mathbf{V} + \ddot{\mathbf{V}}\mathbf{n})\}dA$$

(55)

By carrying out the calculation of the similar nature, the expression for $\langle\sigma\rangle$ for shear flow, as found in [50], is

$$\langle\sigma\rangle = -\bar{P}\mathbf{I} + \begin{pmatrix} \lambda, & 1 + \dfrac{5k+2}{2(k+1)}\phi, & 0 \\[3mm] \dfrac{5k+2}{2(k+1)}\phi, & \lambda + 2\lambda\varepsilon & 0 \\[3mm] 0 & 0 & 0 \end{pmatrix}$$

$$+ \phi\lambda \begin{pmatrix} b_{11} & 0 & 0 \\ 0 & b_{22} & 0 \\ 0 & 0 & b_{33} \end{pmatrix} + \phi\lambda^2 \begin{pmatrix} 0 & F & 0 \\ F & 0 & 0 \\ 0 & 0 & 0 \end{pmatrix}$$

$$(56)$$

where k is the ratio of the viscosities of the disperse and continuous phase, and where

$$\phi = N \cdot \frac{4}{3}\pi a^3 / V$$

and b_{11}, b_{22}, b_{33} and F depend on ε and k in the following manner:

$$(k+1)^2 b_{11} = 4.74k^2 + 1.86k + 2.38$$

$$+ \varepsilon(22.75)k^2 + 11.83k + 8.78)$$

$$(k+1)^3 F = -10.17 + 10.34k - 17.0k^2 - 6.75k^3$$

$$+ \varepsilon(-6.56 + 26.8k + 13.6k^2 + 2.6k^3)$$

$$+ \varepsilon^2(8.8 + 13.57 + 14.2k^2 + 1.17k^3)$$

$$(57)$$

with similar expressions for b_{22} and b_{23}. From (56) we note that the effective viscosity of the suspension is effected by $O(\lambda^2)$ terms. Sun and Jayaraman [50] also compare their normal stress results with experiments and observed that these agree at very low strain rates.

On the experimental side very little work seems to have been carried out. The earlier work in this area, by Krieger, Metzner and Whitlock, Hoffman and others has been carefully documented by Jeffery and Acrivos [13]. In particular Hoffman's [51] result show that, depending upon the volume fraction of the suspended particles, one could find both thinning and thickening behaviour at different shear rates. Thus there is a critical shear rate below which the suspension behaves like a shear thinning fluid which above this value it behaves like a shear thickening fluid. These results indicate that besides understanding the rheological properties of the suspension we need also to understand the structure of the suspension. Moreover the size of the particles plays an important role in deciding the nature of apparent viscosity.

Viscosities of suspended particles in polymeric solutions depend upon dissolved polymer concentration, volume fraction of particles and the shear rate. Usually the relative viscosity is defined to be the ratio of the viscosity of the filled fluid to that of the unfilled fluid at the same shear rate. However recently, it is also being defined as the ratio of the filled to unfilled viscosity at same shear stress rather than at same shear rate. This definition is being preferred because in the latter case, when the relative viscosity is plotted against shear stress asymptotic values of the relative viscosity are obtained. In the case of shear rate the relative viscosity did not, often, reach an asymptotic value at high shear rates. The asymptotic values are correlated with the empirical formulae of the type

$$\eta_r = \exp(\frac{2.5\phi}{1-k\phi})$$

$$\eta_r = (1-\frac{\phi}{\phi_m})^{-2}$$

and so on. Here ϕ is the volume fraction, η_r is the apparent viscosity and k and ϕ_m are some parameters which are determined from experiments.

Acknowledgement: The work reported here has been supported by Grant (A7728) from Natural Sciences and Engineering Research Council of Canada. The authors gratefully acknowledge the support thus received.

160

REFERENCES

1. Stokes, G., On the Effect of the Internal Friction of Fluid on the Motion of Pendulums, *Trans. Camb. Phil. Soc.* vol. 1, p. 8, 1851.

2. Happel, T., and Brenner, H., *Low Reynolds Number Hydrodynamics*, Prentice Hall, Englewood Cliffs, N.J., 1965.

3. Clift, R., Grace, J.R., and Weber, M.E., *Bubbles, Drops and Particles*, Academic Press, New York, 1978.

4. Goldsmith, H., and Mason, S.G., Microrheology of Dispersed Systems, *Rheology*, vol. 4, 85, 1967.

5. Leal, L.G., The Motion of Small Particles in non-Newtonian Fluid, *J. Non-Newtonian Fluid Mech.*, 5, 33, 1979.

6. Caswell, B.,'Sedimentation of Particles in Non-Newtonian Fluids' The Mechanics of Viscoelastic Fluid, *ASME AMD*, 22, 19, 1977.

7. Brunn, P., The Motion of Rigid Particles in Viscoelastic Fluid, *J. Non-Newt. Fluid Mechanics*, 7, 271, 1980.

8. Dairenieh, I.S., and McHugh, A.J., Viscoelastic Fluid Flow Past a Submerged Spheroidal Body, *J. Non-Newtonian Fluid Mech.*, 19, 81, 1985.

9. Chhabra, R.P. in *Encyclopedia of Fluid Mechanics*, ed. N.P. Cheremisinoff, Gulf Publishing, vol. 1, 983, 1986.

10. Einstein A., *Investigations on the Theory of the Brownian Movements*, Dover, N.Y., 1956.

10(a). Batchelor, G.K., Transport Properties of Two-phase Materials with Random Structure, *Ann. Rev. Fluid Mech.*, 6, 227, 1974.

11. Brenner, H., 'Suspension Rheology in *Progressive Heat and Mass Transfer*, vol. 5, Pergamon Press, Oxford, 1972.

12. Jinescu, V.V,, The Rheology of Suspensions, *Int. Chem. Engg*, 14, 397 (1970).

13. Jeffery, D.J., and Acriros, A., The Rheological Properties of Suspension of Rigid Particles, *A.I.Ch.E.J*, 22,417, 1976.

14. Roscoe, R., On the Rheology of a Suspension of Viscoelastic Spheres, *J. Fluid Mech.*, 28, 273, 1966.

15. Goddard, J.D., and Miller, C., Non-linear Effects in the Rheology of Dilute Suspensions, *J. Fluid Mech.*, 28, 657, 1967.

16. Chin, H.B., and Han, C.D., Studies on Droplet Deformation and Breakups I. Droplet Deformation in Extensional Flow. *J. Rheol.*, 23, 557, 1979.

17. Leslie, F.M., The Slow Flow of a Viscoelastic Liquid Past a Sphere, *Quart. J. Mech. and Appl. Math.*, 14, 36, 1961.

18. Oldroyd, J.G., Non-Newtonian Effects in Steady Motion of Some Idealized Elasticoviscous Liquids, *Proc. Roy. Sci. Lond.*, A245, 278, 1958.

19. Giesekus, H. Die simultane Translations - und Rotation sbewgung einer Kugel in iener elasticoviskosen Flüssigkeit. *Rheol. Acta*, 3, 59, 1963.

20. Tiefenbruck, and Leal, L.G., A Numerical Study of the Motion of a Viscoelastic Fluid Past Rigid Spheres and Spherical Bubbles, *J. Non-Newtonian Fluid Mech.*, 10, 115, 1982.

21. Ajayi, O.O., Slow Motion of a Bubble in a Viscoelastic Fluid, *J. Engng. Math.*, 9,273, 1975.

22. Shirotsuka, T., and Kawase, Y., Kagaku, Kogaku, 38, 797, 1974.

23. Phan-Thian, N., and Tanner, R.I., *J. Non-Newtonian Fluid Mech.*, 2, 353, 1977.

24. Johnson, M.W., and Segalman, D., *J. Non-Newtonian Fluid Mech.*, 2, 255, 1977.

25. Hinch, E.J., and Leal, L.G., Effect of Brownian Motion on the Rheological Properties of a Suspension of Non-spherical Particles, *J. Fluid Mech.*, 52, 683, 1972.

26. Frankel, N.A., and Acrisos, A., On the Viscosity of a Concentrated Suspension of Solid Spheres, *Chem. Eng. Sci.*, 22, 847, 1967.

27. Tiefenbruck, G.F., and Leal, L.G., A Note on the Slow Motion of a Bubble in a Viscoelastic Liquid, *J. Non-Newtonian Fluid Mech.*, 7, 251, 1980.

28. Quintana, G.C., Cheh, H.Y., and Madarelli, C.M., The Translation of a Newtonian Droplet in a 4 Constant Oldroyd Fluid, *J. Non-Newtonian Fluid Mechanics*, 22, 253, 1987.

29. Landau, L.D., and Lifshitz, E.M., *Fluid Mechanics*, Addison-Wesley Reading, Mass., 1959.

30. Adachi, K., Yoshioka, N., and Sakai, K., An Investigation of Non-Newtonian Flow Past a Sphere, *J. Non-Newt. Fluid Mech.*, 3, 107, 1978.

31. Zana, E., Tiefenbruck, G., and Leal, L.G., A Note on the Creeping Motion of a Viscoelastic Fluid Past a Sphere, *Rheol. Acta*, 14, 891, 1975.

32. Nakano, Y., and Tien, C., Creeping Flow of a Power Law Fluid Over a Newtonian Sphere, *A. I. Ch. E. J.*, 14, 145, 1968.

33. Acharya A., Mashelkar, R.A., and Ulbrecht, J., Flow of Inelastic and Viscoelastic Fluid Past a Sphere I, II, Rheol. Acta, 15, 454, 471, 1976.

34. Mohan V., and Raghuraman, J., Viscous Fluid of an Ellis Fluid Past a Newtonian Fluid Sphere, Canad. J. Chem. Engng., 54, 228, 1976.

35. Duzhi, G., and Tanner, R.I., The Drag on a Sphere in a Power Law Fluid, J. Non-Newtonian Fluid Mech., 17, 1, 1985.

36. Sugeng, F., and Tanner, R.I., The Drag on a Sphere in a Viscoelastic Fluid with Significant Effects, J. Non-Newt. Fluid Mech., 20, 281, 1986.

37. Mena, B., Manero, O., and Leal, L.G., The Influence of Rheological Properties on the Slow Flow Past Sphere, J. Non-Newt. Fluid Mech., 26, 247, 1987.

38. Chhabra, R.P., Ulherr, P.H.T., and Boger, D.V., The Influence of Fluid Elasticity on the Drag Coefficient for Creeping Flow around a Sphere, J. Non-Newtonian Fluid Mech., 6, 187, 1980.

39. Chilcott, M.D., and Rallison, J.M., Creeping Flow of Dilute-polymer Solutions Past Cylinders and Sphere, J. Non-Newt. Fluid Mech., 29, 381, 1988.

40. Chmielewski, C., Nichols, K.L., and Jayaraman, K., A Comparison of the Drag Coefficients of Spheres Translating in Corn-syrup-based and Polybutene-based Boger Fluid, J. Non-Newt. Fluid Mech., 35, 37, 1990.

41. Barthes-Biesell, D., Motion of a Spherical Macrocapsule Freely Suspended in a Linear Shear Flow, J. Fluid Mech., 100, 831, 1980.

42. Barthes-Biesel, D., and Rallison, J.M. The Time Dependent Deformation of a Capsule Freely Suspended in a Linear Shear Flow, J. Fluid Mech., 113, 251, 1981.

43. Barthes-Biesel, D., and Sgaier, H., Role of Membrane Viscosity in the Orientation and Deformation of a Spherical Capsule Suspended in Shear Flow, J. Fluid Mech., 160, 119, 1985.

44. Batchelor, G.K., The Stress System in a Suspension of Force-free Particles, J. Fluid Mech., 41, 545, 1970.

45. Batchelor, G.K., Developments in Microhydrodynamics, Proc. Theoretical and Appl. Mechanics, ed. W.T. Koiter, North-Holland Publishing Co., 33, 1976.

46. Brenner, H., Rheology of a Dilute Suspension of Axisymmetric Brewnian Particles, Int. J. Multiphase Flow, 1, 195, 1974.

47. Hill, D.A., and Soane, D.S., Rheology of Dispersion of rod-like particles in viscoelastic Fluid, in *Recent Developments in Structual Continua II*, eds. D. DeKee and P.N. Kaloni, Longman, London, 80, 1990.

48. Kaloni, P.N., and Stastna, V., Steady Shear Rheological Behaviour of a Spherical Particles Suspended in a Second Order Fluid, *Poly. Engng. Science*, 23, 465, 1983.

49. Peery, J.H., Fluid Mechanics of Rigid and Deformable Particles in Shear Flow at Low Reynolds Number, Ph.D. Thesis, Princeton University, 1966.

50. Sun, K., and Jayaraman, Bulk Rheology of Dilute Suspensions in Viscoelastic Liquids, *Rheol. Acta*, 23, 84, 1984.

51. Hoffman, R.L. Discontinuous and Dilatant Viscosity behaviour in Concentrated Suspensions I, Observation of Flow Instability. *Trans. Soc. Rheol.*, 16, 155, 1972.

Chapter 8

Rheology of Filled Polymeric Systems

P. J. CARREAU
Centre de Recherche Appliquée Sur les Polymères (CRASP)
Ecole Polytechnique de Montréal
Case Postale 6079, Succursale A
Montréal (Québec), H3C 3A7 Canada

1. INTRODUCTION

In many industrial operations, the fluids or materials handled are non homogeneous, as they consist of more than a single phase. Typical examples include foodstuffs, pharmaceutical products, paints, greases, etc., which are either emulsions or suspensions of solid particles or of liquid droplets. Slurries such as mineral and paper pulps and fermentation broths, biological fluids (such as blood) are other examples of multi-phase systems which have complex rheology. Most of these systems exhibit non-Newtonian behaviour and frequently show time-dependent effects. Many other examples of multi-phase systems can be found in the plastics industry.

1.1 Trends in the plastics industry

Plastics are keeping a fast pace in displacing more and more conventional materials and in creating new opportunities. Technical advantages and financial incentives make polymeric materials an attractive alternative in an increasing number of applications. The development of new resins which, for several decades, concentrated much of the research effort tends to be replaced by new approaches. Government regulations on new products make it now more attractive to rather use existing materials and modify and improve their properties through formulation, blending, etc. The strength of these new tendencies is clearly shown by the rapid emergence of polymer blends and composites as the major market.

Blends and composites as well as high strength engineering plastics play now major roles in many new technologies. In most developing strategic areas, one encounters an increasing demand for high performance plastics and composites. As examples we can cite the growing market in the automobile and aerospace industries for manufacturing bumpers, helicopter blades, etc.

The increase of plastics sales over the recent years is much higher than the increase of the conventional materials. A very large

165

proportion of the market is taken by the polyolefins, low density polyethylene (LDPE), high density polyethylene (HDPE), linear low polyethylene (LLDPE) and finally polypropylene. In fact the only other competing thermoplastics in terms of volume of sales in North America are polystyrene used largely as foam for insulation panels, cups and containers, etc., and polyvinyl chloride (PVC) used for wire coating and window frames and various profiles. The situation in the world is quite similar. The PVC share of the market, however, is different in Europe as PVC is a major material for bottles. The world production of polyolefins in 1980 was estimated to be of 22 Mtons representing some 39 % of the total consumption of plastic materials (see Romanini, 1982). The world consumption of polyolefins increased to some 31 Mtons in 1986 representing then some 57 % of all plastics (Chemical Marketing Reporting, 11/23/87 p. 3, see also Strobel, 1987).

Polyolefin blends attract much interest for technical as well as commercial reasons. For example, about 60 % of all the LLDPE production goes into blends with other polymers such as LDPE, HDPE or ethylene-vinyl acetate copolymers (Hamielec, 1986). The impact of LLDPE and its blends is particularly strong on film blowing where it tends to displace LDPE. In North America, LLDPE represented 43 % of the low density polyethylene market in 1986 while it was only 10 % in Europe. The growth of LLDPE is quite spectacular in North America, as it represented only 13 % of the low density market in 1981.

Even though LLDPE's penetration of the market has been rapid, it has been delayed due to difficult processing. The shift from LDPE to LLDPE often resulted in throughput decreases by 20 to 50 %. This could be avoided by costly equipment modifications, but the most widely used approach is blending LDPE with LLDPE to take advantage of the mechanical properties of the latter and of the good processing characteristics of the former. The blends can usually be processed using conventional equipment (Speed 1982).

Polymers used in commercial blends are often immiscible. Properties of multiphase materials are strongly influenced by the morphology which, in turn, depends on the interactions or chemical affinity of the two polymers, their rheological behaviour and the processing conditions. Considerable research is going on presently worldwide to explore the properties of various blends of commodity thermoplastics and relate material structure, processing, morphology and final properties. Strangely enough, research on blends (at least as far as open literature is concerned) has been lagging behind applications. For example, although blends of polypropylene and LLDPE have been commercially manufactured for more than two decades, it was only recently that technical papers appeared on these blends (Dumoulin et al., 1984, 1987; Yeh and Birley, 1985). Now, considerable literature is available on most combinations of thermoplastics, but most papers are restricted to one or a few aspects of the problems.

If blending represents often a compromise between processing properties and final properties, the use of fillers or additives to thermoplastics to obtain specific properties at a low or rather low cost is an alternative of considerable interest. Filled thermoplastics are rapidly growing industrial materials. Fillers are added to polymers to provide reinforcement to enhance such properties as hardness, high temperature creep resistance, stiffness and opacity of finished products (Nguyen and Ishida, 1987; Menendez and White, 1984). Fillers, however, play an important role on the processing characteristics. Addition of fillers increases the pressure drop across extrusion dies due to an increase in melt viscosity. It also gives rise to less die swell suggesting that the elasticity of the material is decreased. Hence, the study of the rheological properties of filled polymeric materials is of prime importance for solving the different problems related to polymer composite processing.

1.2 Scope of this chapter

In predicting the flow behaviour of polymer composites and blends, considerable difficulties arise from the state of interface (Bigg, 1982)), from the variation of that state with time and processing, orientation of solid particles, agglomeration, breakage or coalescence, etc., under flow conditions. Strictly speaking, the rheology of multi-phase systems is a non-sense. One should rather try to solve problems using equations of fluid mechanics for two-phase flow. For many systems, including viscoelastic polymeric materials, it is more practical, however, to define rheological properties for the mixture and eventually account for changes due to processing.

The rheology of multi-phase polymeric systems is a very complex subject of which the principles are not yet well defined. Appropriate constitutive relations for even the most simple model systems (for example suspensions of glass beads in a polymer matrix) still need to be developed and many published rheological results are debatable. In section 2, we review the basic principles of viscosity measurement using capillary rheometers and stress difficulties encountered when dealing with multi-phase systems. In section 3, we discuss key results of the literature and present some recent results to illustrate the difficulties of the subject. This is followed by brief concluding remarks. Obviously, this chapter is far from being exhaustive and the reader interested in learning more should refer to the excellent review done by Metzner (1985) on the rheology of filled systems and to the more recent and comprehensive review on the rheology of multi-phase systems by Utracki (1988).

2. RHEOLOGICAL MEASUREMENTS

The major methods for measuring rheological properties are discussed in many textbooks, including that of Dealy (1982) and the recent book edited by Collyer and Clegg (1988). We will, however, briefly review the concepts of measurements using capillaries. Capillary measurements may be highly informative on the feasibility of using continuum concepts to describe multi-phase systems.

2.1 Rabinowitsch's analysis

Capillary viscosity measurements are based on relationships between pressure drop across the capillary length and the flow rate. This is not a direct measurement and one must first choose a constitutive relation for the fluid (Newtonian behaviour, power-law, Ellis model, etc.). It is possible to avoid this by following the Rabinowitsch procedure (see Bird et al., 1960).

The approach assumes the following:

laminar flow

steady state: $\dfrac{\partial}{\partial t} = 0$

fully developed unidirectional flow:

$V_r = V_\theta = 0$

$V_z = V_z \ (r)$

no slip at the capillary walls:

$V_z \ (r = R) = 0$

and $\quad \dfrac{dV_z}{dr} = \dfrac{dV_z}{dr} \ (\tau_{rz}) \qquad$ or

$\eta = \eta \ (\tau_{rz})$, the viscosity is a unique function of the shear rate or shear stress.

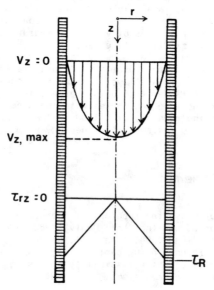

FIGURE 2.1. Flow in a capillary.

The volumetric flow rate is expressed by:

$$Q = 2\pi \int_0^R V_z \, r \, dr \tag{2.1-1}$$

or integrating by parts:

$$Q = \pi V_z r^2 \Big]_0^R - \int_0^R \pi \left(\frac{dV_z}{dr}\right) r^2 \, dr \tag{2.1-2}$$

For non-slip conditions at the capillary walls, the integrated term is equal to zero. On the other hand, a momentum balance on differential element $2\pi r \Delta r L$ yields:

$$\boxed{\tau_{rz} = \frac{r}{R} \, \tau_R} \tag{2.1-3}$$

where: $\qquad\qquad \tau_R = \dfrac{\Delta PR}{2L}$, shear stress at the wall

ΔP is the total pressure drop. Equation (2.1-2) becomes after the following change of variable, $r \rightarrow \tau_{rz}$

$$\frac{Q}{\pi R^3} = \frac{1}{\tau_R^3} \int_0^{\tau_R} \tau_{rz}^2 \left(-\frac{dV_z}{dr}\right) d\tau_{rz} \tag{2.1-4}$$

Taking the derivative of this expression with respect to τ_R yields:

$$\frac{d}{d\tau_R} \left(\frac{\tau_R^3 Q}{\pi R^3}\right) = \frac{d}{d\tau_R} \int_0^{\tau_R} \tau_{rz}^2 \left(-\frac{dV_z}{dr}\right) d\tau_{rz} \tag{2.1-5}$$

Using Leibnitz's rule

$$= \left\{ \tau_{rz}^2 \left(-\frac{dV_z}{dr}\right) \right\} \Big|_{\tau_R}$$

Hence the shear rate evaluated at the wall is expressed by:

$$-\dot{\gamma}_R = \left(-\frac{dV_z}{dr}\right) \Big|_{r=R} = \frac{1}{\pi R^3 \tau_R^2} \cdot \frac{d}{d\tau_R} (\tau_R^3 Q) \tag{2.1-6}$$

This result can now be used to obtain the viscosity from basic measurements of Q et ΔP.

For example, at τ_{R1}, the slope is p_1, hence:

$$\left(-\frac{dV_z}{dr}\right)_1 \Big|_{r=R} = -\dot{\gamma}_{R1} = \frac{1}{\pi R^3 \tau_{R1}^2} \, p_1$$

and the corresponding viscosity is:

$$\eta(\dot{\gamma}_{R1}) = - \frac{\tau_{R1}}{\dot{\gamma}_{R1}} = \frac{\pi R^3 \tau_{R1}^3}{p_1}$$

In practice, it is not easy to determine with accuracy a slope from graphical data. We rather prefer to fit a monotonous simple polynomial expression to result (2.1-6), re-written in the following form:

$$-\dot{\gamma}_R = \frac{1}{4\tau_R^2} \cdot \frac{d}{d\tau_R} (\tau_R^3 \frac{4Q}{\pi R^3})$$

$$= \frac{3}{4} (\frac{4Q}{\pi R^3}) + \frac{1}{4} (\frac{4Q}{\pi R^3}) \cdot \frac{d\ln (4 Q/\pi R^3)}{d\ln \tau_R} \qquad (2.1-7)$$

Defining

$$n' = \frac{d\ln \tau_R}{d\ln (4Q/\pi R^3)} \qquad (2.1-8)$$

Equation 2.1-8 can be written as:

$$-\dot{\gamma}_R = \frac{3n' + 1}{4n'} (\frac{4Q}{\pi R^3}) \qquad (2.1-9)$$

The wall shear rate can be obtained directly from the rheogram τ_R vs. $4Q/\pi R^3$ on log-log scales. The curve is frequently close to a straight line, and the slope, n', is then a constant equal to the power-law index. Otherwise, the curve can be linearized parts by parts or a polynomial used to fit the data. An equivalent approach is to calculate the apparent viscosity from:

$$\eta_a = -\tau_R/\dot{\gamma}_a = -(\frac{\Delta PR}{2L})/(\frac{4Q}{\pi R^3}) \qquad (2.1-10)$$

This is the correct expression for the viscosity of a Newtonian fluid. A log-log plot of the apparent viscosity for a high density polyethylene (HDPE 16A of DuPont Canada) at 180°C is shown in Figure 2.2 (data shown without correction). The slope is equal to n'-1, but it is not constant over the whole range of shear rate. At high shear rate, $\dot{\gamma}_a >$ 100 s^{-1}, the slope is constant and n' is equal to n (power-law region). Knowing n', the true shear rate at the wall can now be calculated from Eq. (2.1-9) and the non-Newtonian viscosity at the wall shear rate is by definition expressed by:

$$\eta (\dot{\gamma}_R) = - \frac{\tau_R}{\dot{\gamma}_R} \qquad (2.1-11)$$

170

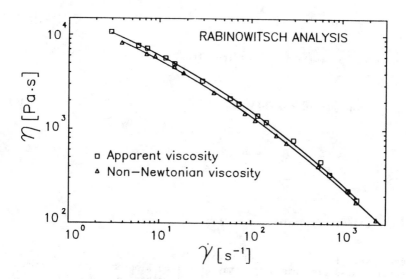

FIGURE 2.2. Apparent and non-Newtonian viscosities of a high density polyethylene melt (HDPE 16A of DuPont) at 180°C.

We note that the corrected viscosity is slightly below the apparent viscosity curve, but since log-log representation is used, the Rabinowitsch correction is not negligible.

Comments:

For a power-law fluid:

$$\tau_R = m \mid \dot{\gamma}_R \mid^n \tag{2.1-12}$$

and $\ln \tau_R = \ln m + n \ln |\dot{\gamma}_R|$

Combining with Equation (2.1-9), we get:

$$\ln \tau_R = \ln m + n \ln \left(\frac{3n'+1}{4n'}\right) + n \ln \left(\frac{4Q}{\pi R^3}\right) \tag{2.1-13}$$

Hence $$\frac{d \ln \tau_R}{d \ln (4Q/\pi R^3)} = n = n' \tag{2.1-14}$$

and the intercept of $\ln \tau_R$ vs. $\ln (4Q/\pi R^3)$ is:

171

$$\ln K' = \ln m + \ln \left(\frac{3n + 1}{4n}\right) \qquad (2.1\text{-}15)$$

The power-law parameters are obtained from:

$$n = n' = d \ln \tau_R / d \ln (4Q/\pi R^3) \qquad (2.1\text{-}16)$$

and
$$\boxed{m = K' \left(\frac{4n}{3n + 1}\right)^n} \qquad (2.1\text{-}17)$$

2.2 End effects or Bagley correction

Unless a very long capillary is used, L/D > 100, end effects (entrance and exit excess pressure drops) may affect considerably the accuracy of the determined viscosity. The empirical method proposed by Bagley (1957) can be used to correct for end effects and also obtain an estimation of the fluid's elasticity. The method consists of determining the excess pressure drops at the entrance and exit of the capillary. A typical pressure profile is illustrated in Figure 2.3.

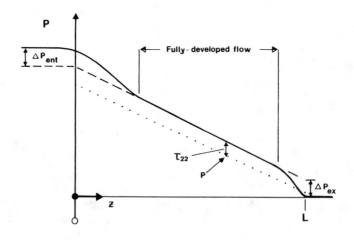

FIGURE 2.3. Pressure profile along the capillary axis.

The shear stress at the wall is corrected by

$$\tau_R = \frac{\Delta PR}{2(L + e_0R)}$$

(2.2-1)

This concept is useful provided that the end correction factor, e_0, is independent of the capillary geometry, i.e. the capillary has to be long enough that fully developed flow is attained before the fluid exits the capillary. The parameter e_0 is determined by constructing the curves of ΔP vs. L/R at constant shear rate or $(4Q/\pi R^3)$. The procedure is illustrated in Figure 2.4 for HDPE data obtained with four different capillaries on an Instron rheometer. For four different shear rates, the applied load or pressure in the reservoir is plotted as a function of L/R. Straight lines are obtained and extrapolating to the axis, we obtain values of e_0 which vary from 4 to 6.4. Then Eq. (2.2-1) can be used to calculate a corrected shear stress and the viscosity is re-calculated using Eq. (2.1-11). These corrections are rather important as shown in Figure 2.5 for two sets of data. Clearly, the uncorrected viscosity data obtained with the capillary of L/D equal to 5 show very large departures from the corrected data.

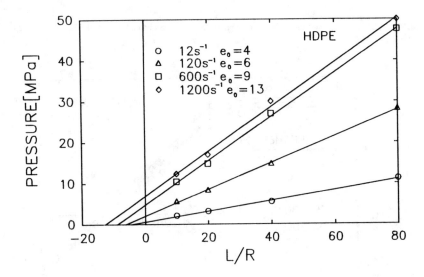

FIGURE 2.4. Bagley corrections for a high density polyethylene at 180°C (DuPont 16A).

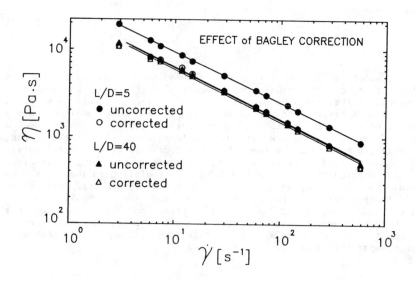

FIGURE 2.5. Uncorrected and corrected viscosities of HDPE 16A of DuPont at 180°C.

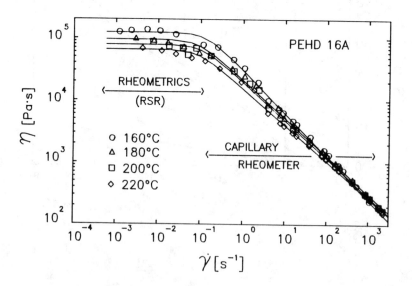

FIGURE 2.6. Shear viscosity as a function of shear rate for HDPE 16A at different temperatures.

In Figure 2.6 we report the corrected viscosity data for HDPE 16A obtained at four different temperatures and using three different rheometers. The low shear rate data were obtained on a constant stress rheometer (Rheometrics RSR). The medium range shear rate data were obtained from a gas driven capillary (C.I.L.) viscometer) and the high shear rate data from an Instron capillary rheometer. Although there is virtually no superposition, the continuity between the three sets of data is good. Notice that over six decades of shear rate are covered by these data.

Finally in Figure 2.7, we report Bagley's plots for a polypropylene reinforced with 20% (per weight) of glass fibers. Obviously, the applied load (or reservoir pressure) is not a linear function of L/D at most values of the piston speed. As many effects (fibers orientations or/and migration, viscous dissipation, pressure dependence) may be responsible for the non-linearity. The extrapolation to obtain e_o is not valid. Bagley's corrections for this filled polymeric system are meaningless and uncorrected data should be reported for the viscosity which is clearly dependent on the hydrodynamics and geometry of the capillary system. Similar results have been obtained for other glass fiber reinforced polymeric systems.

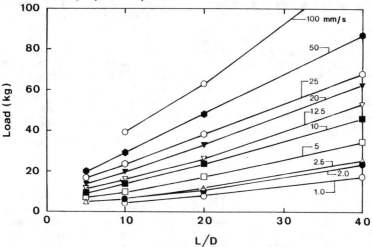

FIGURE 2.7. Bagley's corrections for a polypropylene filled with 20% (per weight) of glass fibers at 180°C, using an Instron capillary rheometer. The parameter is the piston speed.

2.3 Mooney correction

The Rabinowitsch method allows for the determination of the shear rate at the wall provided the rheology of the fluid is homogeneous, i.e. the viscosity is a unique function of the shear rate or shear stress. This is not necessarily the case with multi-phase systems. For example, due to the finite dimensions of the filler particles, special wall effects resulting in an apparent slip may be observed. Other possible effects include migration and orientation of particles, coalescence or phase inversion in blends, etc. These effects may be detected and eventually corrected by using the following method.

Let us suppose that any wall effects can be modelled as a slip velocity at the wall of the capillary. This velocity can be positive (equivalent to a dilution effect) or negative (equivalent to an adsorption or concentration effect). Result (2.1-2) becomes:

$$\frac{Q}{\pi R^3} = \frac{V_s}{R} + \frac{1}{\tau_R^3} \int_0^{\tau_R} \tau_{rz}^2 \left(-\frac{dV_z}{dr}\right) d\tau_{rz} \tag{2.3-1}$$

where V_s is the apparent slip velocity. This result can be re-written as:

$$\frac{Q}{\pi R^3 \tau_R} = \frac{\beta}{R} + \frac{1}{\tau_R^4} \int_0^{\tau_R} \tau_{rz}^2 \left(-\frac{dV_z}{dr}\right) d\tau_{rz} \tag{2.3-2}$$

where β is the apparent slip coefficient. We note that the integral on the right inside of this equation is independent of the capillary radius. It is a unique function of the shear stress (or shear rate) evaluated at the wall. Hence, to ascertain if there are wall effects, the Bagley corrections are first determined using a set of different L/R capillaries but of the same radius. Then another series of experiments for capillaries of different radius is conducted and plots of $(Q/\pi R^3 \tau_R)$ vs. τ_R (corrected values) are made for each R. In absence of wall effects, β is equal to zero and a single curve should be obtained. If a set of different curves is obtained, the value of β can be easily determined by plotting $Q/\pi R^3 \tau_R$ vs. $1/R$ at a given shear stress.

Comments:

In obtaining the rheology of multi-phase systems from capillary rheometers, one should look for other possible effects which may limit considerably the usefulness of the data. The most important other sources of difficulties are the heat generated by viscous dissipation and the influence of pressure on viscosity. For polymer melts, these effects may be very important and render the measurements useless.

3. RHEOLOGY OF FILLED THERMOPLASTICS

Excellent brief reviews of the rheology of filled systems can be found in Metzner (1985) and in Kamal and Mutel (1986); an extensive review has been presented by Utracki (1988). However, the papers written by Poslinski et al. (1988) are excellent starting points for the understanding of the effects of inert fillers on the rheological properties of polymers.

The rheological properties of thermoplastics depend on the volume or mass fraction of the filler, the geometry of the filler particles, their orientation and distribution in the polymer matrix, and finally on the particle-particle and particle-matrix interactions.

3.1 Viscosity of suspensions of spheres

The viscosity of low interaction spherical particles in Newtonian systems is illustrated in Figure 3.1, adapted from Thomas (1965).

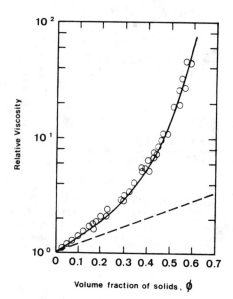

FIGURE 3.1. Relative viscosity vs. concentration for suspensions of spheres of low interaction - narrow size distributions - particle diameters in the range of 0.1 to 440 μm (from Thomas, 1965).
————— Equation 3.1-1
- - - - - - - - Equation 3.1-2

It is interesting to note that the relative viscosity is a unique function of the volumetric fraction of solids. The whole curve can be described by the Maron and Pierce (1956) empirical equation:

$$\mu_r = \mu/\mu_m = [1 - \varphi/\varphi_m]^{-2} \tag{3.1-1}$$

where φ_m is the solid fraction at maximum packing, equal to 0.68 for solid spheres. Also shown in the figure is the theoretical result of Einstein (1906; 1911) obtained for very dilute systems:

$$\mu_r = \mu/\mu_m = [1 + \frac{5}{2}\varphi] \tag{3.1-2}$$

The non-Newtonian behaviour of polymers filled with low interacting spheres is very similar to that of non filled polymers, at least up to a solid fraction close to maximum packing. Figure 3.2 shows the data obtained by Poslinski et al. (1988) for filled thermoplastics and the description of the data using the Carreau (1972) equation:

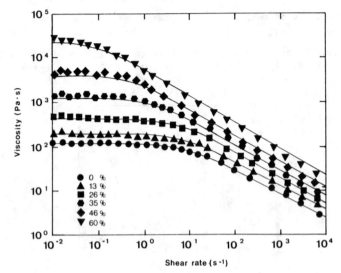

FIGURE 3.2. Shear viscosity of glass spheres dispersed in a thermoplastic polymer at 150°C. Data from Poslinski et al. (1988). The glass spheres average diameter was 15 μm with a narrow size distribution. The solid lines are the predictions using the Carreau equation.

$$\eta_c = \eta_{co}[1 + (\lambda\dot{\gamma})^2]^{(n-1)/2} \qquad (3.1-3)$$

where the subscript c refers to the composite and η_{co} is the zero-shear viscosity. We note that the data of Poslinski et al. (1988) were obtained by using a cone-and-plate rheometer (Rheometrics RMS) for the lower shear rates and an Instron capillary viscometer for high shear rates. The Mooney correction was found to be negligible and the excellent agreement in both sets of data tends to prove that these filled systems behave like homogeneous fluids. The time constant for the composite is given by a Maron-Pierce equation, i.e.:

$$\lambda_r = \lambda_c/\lambda_m = [1 - \varphi/\varphi_m]^{-2} \qquad (3.1-4)$$

Notice the large increases of the reduced time constant, up to three decades, with increasing solid content. This suggests that, contrarily to the usual thinking, the addition of solid particles increases the fluid's elasticity.

Another very important finding of Poslinski et al. (1988) is that Eq. (3.1-1) can be applied to non-Newtonian polymeric systems, provided that the viscosity values of the composite and of the matrix are compared at the same shear stress, i.e. Eq.(3.1-1) has to be re-written in the following form:

$$\eta_r = (\eta_c/\eta_m)\big|_{\tau_{12}} = [1 - \varphi/\varphi_m]^{-2} \tag{3.1-5}$$

This is a very important result. It means that the viscosity of thermo-plastics filled with narrow size distribution and low interacting spheres can be predicted from the viscosity of the matrix. In fact the findings of Poslinski et al. (1988) have much wider implications as we will see below.

It has been observed that polydispersity reduces the viscosity of filled systems at fixed solid concentration (see Eveson, 1959). Polinski et al. (1988) have shown that a formula for the maximum packing parameter proposed by Gupta and Seshadri (1986) can be used to calculate with the help of Eq. (3.1-5) the relative viscosity of polymers filled with polydisperse spheres. Equation (3.1-5) was found to give very good predictions for the viscosity of bimodal systems (Poslinski et al., 1988). At very high solids fractions ($\varphi = 0.60$), the reduced viscosity goes through a very low minimum for systems containing approximately 15 % of 15 μm glass beads (φ_s), the rest consisting of 78 μm beads. The viscosity reduction compared to the monodispersed systems is of the order of ten. This is non negligible for engineering applications and larger viscosity reductions have been reported. Finally, these results suggest that φ_m can be taken as a fitting parameter for polymer systems filled with particles of more complex shapes.

3.2 Elasticity of suspensions of spheres

There are so far limited data on the elastic properties of suspensions (normal stresses in steady simple shear flow, storage modulus, etc...). Both the Bagley correction term and the extrudate swell are known to decrease with solids content. On the other hand, the (elastic) storage modulus was found by Faulkner and Schmidt (1977) to increase with the solids fraction according to the formula (for solids concentration less than 0.26):

$$G'_r \equiv G'_c/G'_m = 1.0 + 1.8\varphi \tag{3.2-1}$$

whereas the (viscous) loss modulus follows the relation:

$$G''_r \equiv G''_c/G''_m = 1.0 + 2.0\varphi + 3.3\varphi^2 \tag{3.2-2}$$

As the viscous forces are increasing more rapidly than the elastic forces, this result is in agreement with the observations on extrudate swell and entrance effects. The recent results of Polinski et al. (1988) are somewhat in contradiction with those of Faulkner and Schmidt. They found that the storage modulus (for frequencies larger than 10 rad/s) can be correlated by a Maron-Pierce type equation, i.e.:

$$G'_r = [1 - \varphi/\varphi_m]^{-2} \tag{3.2-3}$$

179

where the moduli are obtained at the same frequency. The loss modulus follows a different correlation:

$$G''_r = \left[1 + \frac{0.75\varphi/\varphi_m}{1 - \varphi/\varphi_m} \right]^2 \qquad (3.2-4)$$

These results indicate that the elastic forces increase more rapidly than the viscous forces. They also measured the primary normal stress differences in steady simple shear flow. Again the Maron-Pierce type equation is successful in correlating the data:

$$\psi_{1r} = \psi_{1c}/\psi_{1m} = [1 - \varphi/\varphi_m]^{-2} \qquad (3.2-5)$$

where the primary normal stress coefficient, $\psi_1 = -(\tau_{11} - \tau_{22})/\dot{\gamma}^2$, is compared at the same shear rate, $\dot{\gamma}$.

It is obvious that more research work is needed to clarify how fillers affect the elastic properties of the polymeric matrix. Clearly, the general impression that elasticity is reduced as the solids concentration increases is not necessarily always true.

3.3 Effects of Particle Geometry

The effects of particle geometry on the viscosity of filled thermoplastics have been reported by Kitano et al. (1981). Their results (obtained in the low shear rate region) show that the viscosity increases much more rapidly for particles of large aspect ratio (fibers). According to Kitano et al., Eq. (3.1-5) can still be used to adequately describe the data by using a single parameter, φ_m, the maximum packing. Obviously, Eq. (3.1-5) is a very useful empiricism, but it should be used with caution, especially if the particles interact and possibly form aggregates. Also, in the case of fibers, orientation due to blending and processing could possibly affect appreciably the viscosity as well as the other rheological properties.

Difficulties encountered in measuring the viscosity of fiber reinforced thermoplastics are stressed in Figure 3.3 which reports the shear viscosity at 180°C of a polypropylene filled with 20% of glass fibers. The Bagley plots for this system are shown in Figure 2.7. Six different rheometers were used to obtain these data (three at Ecole Polytechnique of Montreal: Rheometrics RSR, Bohlin VOR-HTU and Instron capillary rheometers; three at the CEMEF of Ecole des Mines de Paris, Sophia Antipolis, FRANCE, Instron cone-and-plate, Instron capillary and Rheoplast capillary rheometers). The rotating rheometers were used to determine the viscosity at low shear rates. The samples were initially pre-sheared to orient the fibers in the direction of the flow field. The high shear-rate data were obtained with the capillary viscometers. The data obtained from both Instron capillary rheometers are in agreement, but considerably higher than the values obtained with the Rheoplast instrument which allows for preshearing before injection in the capillary. These results and others are discussed in a forthcoming publication (Ausias et al., 1990).

180

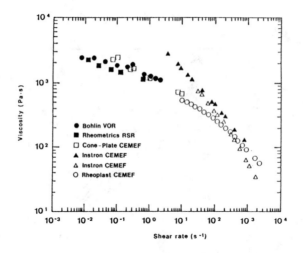

FIGURE 3.3. Shear viscosity of a polypropylene filled with 20% (weight) of glass fibers.

Probably due to orientation in shear flow, Eq. (3.1-1) considerably overpredicts the viscosity of thermoplastics filled with long fibers (aspect ratio > 30). The results of Chan et al. (1978) are of considerable interest as they show that the reduced viscosity at a given shear rate increases much more rapidly with the glass fibers content in a case of a high density polyethylene flow than for a polystyrene. In some cases, the viscosity of filled systems is barely higher than that of the matrix, and in rare cases even lower (see next section).

3.4 Interactive particles

When interparticle interactions are important compared to viscous forces, many complications can arise. A major one is the agglomeration of particles which results in a major increase of the viscosity, mainly at low shear rates, as shown by Nielsen (1977) for glass beads in presence of water in a hydrocarbon liquid. Not only the behaviour becomes non-Newtonian, but the suspension appears to show a yield stress (intercept on the shear stress vs. shear rate curve).

Another interesting example is reported by Minagawa and White (1975) for a high density polyethylene filled with rutile particles. The increases in viscosity at low shear stress for high solid loading was found to be quite spectacular. For the highly concentrated suspensions, the low shear rate or low shear stress viscosity appeared to be unbounded. This is an indication of a yield stress. The viscosity of these systems depends strongly on the interfacial properties and sizes of the particles. Also, the blending conditions may alter the viscosity of the systems.

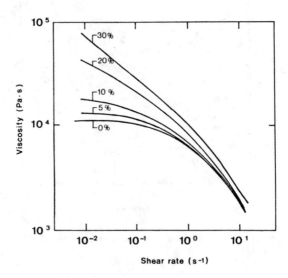

FIGURE 3.4. Viscosity of a high molecular weight polystyrene filled with carbon black (24 m²/g) at 180°C. Data from Lakdawala and Salovey (1987).

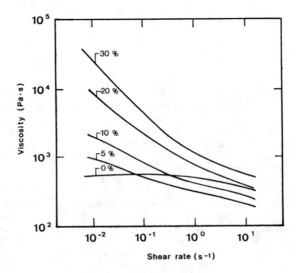

FIGURE 3.5. Viscosity of a high molecular weight polystyrene filled with carbon black (24 m²/g) at 230°C. Data from Lakdawala and Salovey (1987).

Carbon black particles are known to yield very different rheologies dependent on the properties and sizes of the particles. Figure 3.4 and 3.5 show two sets of viscosity data reported by Lakdawala and Salovey (1987) for a polystyrene melt filled with carbon black. Notice the apparently unbounded viscosity at low shear rate for high filler content. The behaviour at 230°C is considerably different from that at 180°C, mainly for low solids systems which exhibit viscosities even lower than the unfilled matrix. This is not uncommon and similar results were obtained in our laboratory for a high density polyethylene.

Figure 3.6 shows the relative viscosity of the polyethylene filled with five different carbon blacks. The viscosity data are compared at $\dot\gamma = 10$ s^{-1} and the solids concentration is on a weight basis. The different carbon blacks were obtained from two suppliers and their main characteristics can be found in Dufresne (1989). Obviously, the viscosity is strongly dependent on the surface properties of the carbon black. Properties of these systems have been discussed by Malik et al. (1988).

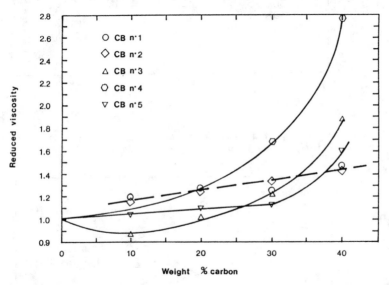

FIGURE 3.6. Relative viscosity of a high density polyethylene filled with carbon black particles at $\dot\gamma = 10$ s^{-1} (data from Dufresne, 1989).

To stress again the complexity of the rheology of filled polymeric systems, we report in the following three figures viscosity and normal stress data recently obtained for polyethylene oxide (PEO) solutions with and without the addition of glass beads. The molecular weight of the PEO was 5.10^6 and three polymer concentrations (1, 2 and 3% per weight) in a mixture containing 50% (per weight) water and 50% glycerol were used. Untreated glass beads with an average diameter of 35 μm and narrow size distribution were used. All data reported here were obtained using a Weissenberg rheogoniometer with the cone-and-plate geometry (2° cone). The validity of the low shear-rate data was confirmed by using a Rheometric RSR instrument.

Figure 3.7 shows the influence of the glass beads loading on the viscosity of the 1% PEO solution. The solid line is the fit of the Carreau equation for the unfilled solution. The curve is typical for polymer solutions. When glass beads are added to the solution, the viscosity is shown to decrease up to a solids content equal to 30% (per weight). The decrease is more pronounced in the low shear rate region. As loading is increased to 40%, the viscosity is then considerably increased.

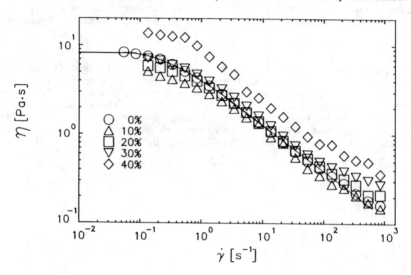

FIGURE 3.7. Viscosity of a 1% polyethylene oxide solution in 50%/50% water/glycerol for various loading of glass beads at 25°C.

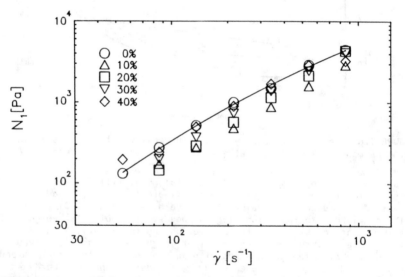

FIGURE 3.8. Primary normal stress differences for a solution in 50%/50% water/glycerol for various loading of glass beads at 25°C.

184

Figure 3.8 shows the primary normal stress differences for the 1% poly-ethylene oxide solutions. The pattern is similar to that observed for the viscosity, i.e. the primary normal stress differences are decreased with the addition of glass beads. The lower values are obtained for the 10% loading. At 40% loading, the normal stress increases to the values obtained for the unfilled solution. These results indicate that the elasticity of these suspensions decreases with the solids content. This is in contradiction with findings of Poslinski et al. (1988) dis-cussed above. A similar behaviour was observed for the 2% and 3% PEO solutions.

Clearly the glass beads are interactive and the reduction of viscosity can be attributed to the following phenomena. The polymer can adsorb on the surface of the glass beads resulting in a somewhat depleted solution compared to the unfilled solution, as postulated by Otsubo (1986). The expected polymer concentration reduction is of the order of a few mg/L and this is far from being large enough to explain the viscosity reduction for solution containing 1 to 3% of polymer. The specific viscosity defined by

$$\eta_{sp} = \frac{\eta_{susp} - \eta_{sol}}{\eta_{sol}} \quad , \qquad\qquad (3.4\text{-}1)$$

where η_{susp} is the viscosity of the filled polymer solution and η_{sol} is the viscosity of the unfilled solution, is plotted in Figure 3.9 for the three polymer concentrations.

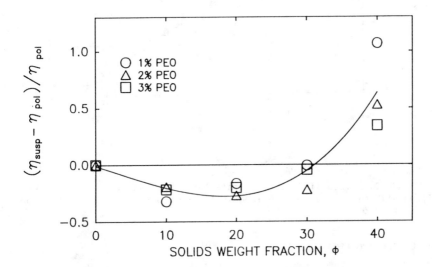

FIGURE 3.9. Specific viscosity for glass beads suspensions in poly-ethylene oxide solutions at a shear rate equal to 0.34 s⁻¹. The solid line is approximately the average for the three polymer concentrations.

It is interesting to note from Figure 3.9 that the specific viscosity is approximately independent of the polymer concentration at least for loadings up to 20% in glass beads. The viscosity reduction at low shear rate (0.34 s⁻¹) is of the order of 25% and can be explained only by the influence of the glass beads on the polymeric chain conformation and entanglements (see for a discussion on specific interactions Lara and Schreiber, 1990).

Finally, we report some of our recent data on mica-polyethylene composites. Figure 3.10 shows the viscosity as a function of shear rate for different concentrations of mica (weight percent). The viscosity data obtained on the Rheometrics RSR at low shear rate are consistent with those obtained on the capillary rheometer. Due to alignment and orientation of the mica flakes, the viscosity increases for the highly concentrated suspensions are not large. At comparable loading, the viscosity increase with mica is considerably less than with carbon black.

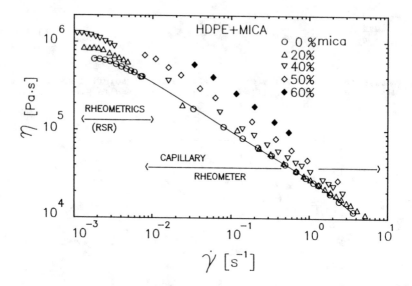

FIGURE 3.10. Viscosity vs. shear rate for a high density polyethylene filled with mica. The polyethylene is the same as for results presented in Figure 3.6.

4) CONCLUDING REMARKS

The rheology of high molecular weight polymers is very complex: under simple shear conditions, the viscosity is shear-thinning (decreases of four decades with shear rate are not uncommon) and important normal stresses are exhibited; in transient flows, the polymer's elasticity gives rise to time-dependent effects coupled with operating conditions which are very difficult to predict; in more complex flow situations where the extensional component is important, elongational or extensional viscosity is likely to play a significant role. Shear viscosity may be of little utility in correlating effects observed in complex flows. Considerable research efforts are currently devoted to the clarification of basic phenomena observed with polymer melts and solutions.

Obviously, the rheology of polymeric multiphase systems is far from being understood. The only aspect that appears to be well established is the shear viscosity of suspensions of non interacting (or of low interactions) spheres. In that respect, Eq. (3.1-5) combined with viscosity equations such as Eq. (3.1-3) is a very useful empiricism to predict the viscosity of filled thermoplastics. For large aspect ratio particles, it should be used with caution. Quite powerful theories are also available to predict the viscosity of dilute suspensions of rigid rods. On the other hand, the viscosity of suspensions of interactive particles depends strongly on the surface properties of the particles, size, geometry and blending methods. Results in the literature should be used with extreme care if one wishes to predict the viscosity of such systems.

The elasticity of filled thermoplastics remains a subject of considerable controversy. The elastic forces may increase slower or faster than the viscous forces in loaded systems. The effects of fillers on the transient response and on the extensional viscosity of thermoplastics remain open questions. Definitely, more research is needed in order to clarify these crucial points.

5. ACKNOWLEDGEMENTS

The new rheological data presented in this chapter have been obtained by my colleagues, associates and students: G. Ausias, A. Dufresne, P.A. Lavoie, T. Malik, Chunk Tu Wong. I thank them all for their most appreciated contribution.

6. NOMENCLATURE

D	Diameter of capillary, m
e_o	Bagley's entrance correction, defined by Eq. (2.2-1)
G'_c, G'_m	Storage modulus of composite, of matrix, Pa
G''_c, G''_m	Loss modulus of composite, of matrix, Pa
G'_r	Reduced storage modulus, defined by Eq. (3.2-1)
G''_r	Reduced loss modulus, defined by Eq. (3.2-2)
L	Length of capillary, m
m	Power-law parameter, Eq. (2.1-12), Pa.s^n
n	Power-law index, Eq. (2.1-12)
n'	Slope defined by Eq. (2.1-8)
N_1	Primary normal stress difference = $-(\tau_{11}-\tau_{22})$, Pa
p	Slope of $\tau_R^3 Q$ vs. τ_R
P	Pressure, Pa

Q	Flow rate, m^3/s
r	Radial position, m
R	Radius of capillary, m
V_r, V_o, V_z	Velocity components in cylindrical coordinates, m/s
V_s	Slip velocity at capillary walls, m/s

Greek letters

β	Slip coefficient $= V_s/\tau_R$, m/(s.Pa)
$\dot{\gamma}$	Shear rate, s^{-1}
$\dot{\gamma}_a$	Apparent shear rate $= 4Q/(\pi R^3)$, s^{-1}
$\dot{\gamma}_R$	Shear rate evaluated at the capillary walls, s^{-1}
η	Non-Newtonian viscosity, Pa.s
η_a	Apparent viscosity defined by Eq. (2.1-10), Pa.s
η_c	Viscosity of composite, Pa.s
η_{co}	Zero-shear viscosity of composite, Pa.s
η_m	Viscosity of matrix, Pa.s
η_r	Reduced viscosity
η_{sp}	Specific viscosity defined by Eq. (3.4-1)
λ	Time constant in Carreau's equation (3.1-3), s
λ_c, λ_m	Time constant of composite, of matrix, s
λ_r	Reduced time constant defined by Eq. (3.1-4)
μ	Newtonian viscosity, Pa.s
μ_m	Newtonian viscosity of matrix, Pa.s
μ_r	Relative or reduced viscosity defined by Eq. (3.1-2)
τ_{12}, τ_{rz}	Shear stress, Pa
$-(\tau_{11}-\tau_{22})$	Primary normal stress difference, Pa
φ	Fraction of solids
φ_m	Fraction of solids at maximum packing
ψ_1	Primary normal stress coefficient $= -(\tau_{11}-\tau_{22})/\dot{\gamma}^2$, $Pa.s^2$
ψ_{1c}, ψ_{1m}	Primary normal stress coefficient of composite, of matrix, $Pa.s^2$
ψ_{1r}	Reduced primary normal stress coefficient defined by Eq. 3.2-5.

7. REFERENCES

Ausias, G., Agassant, J.F., Vincent, M., Carreau, P.J., and Lafleur, P.G., "Rheological Properties of Polypropylene Reinforced with Glass Fibers", in preparation (1990).

Bagley, E.B., "End Corrections in the Capillary Flow of Polyethylene", J. Appl. Phys., 28, 624, 1957.

Bigg, D.M., "Rheological Analysis of Highly Loaded Polymeric Composites filled with Non-agglomerating Spherical Filler Particles", Polym. Eng. Sci., 22, 512, 1982.

Bird, R.B., Stewart, W.E., Ligthfoot, E.N., "Transport Phenomena", John Wiley, New York, 1960.

Carreau, P.J., "Rheological Equations from Molecular Network Theories", Trans. Soc. Rheol, 16, 99, 1972.

Collyer, A.A., and Clegg, D.W., "Rheological Measurement", Elsevier Applied Science, London and New York, 1982.

Chan, Y., White, J.L., and Oyanagi, Y., "A Fundamental Study of the Rheological Properties of Glass-Fiber-Reinforced Polyethylene and Polystyrene Melts", J. Rheol., 22, 507, 1978.

Dealy, J.M., "Rheometers for Molten Plastics", Van Nostrand Reinhold, New York, 1982.

Dufresne, A., "Propriétés mécaniques, électriques et rhéologiques de matériaux thermoplastiques renforcés de noir de carbone, M.A.Sc. thesis, Ecole Polytechnique of Montreal, 1989.

Dumoulin, M.M., Farha, C., and Utracki, L.A., "Rheological and Mechanical Properties of Ternary Blends of Linear-Low-Density Polyethylene/Polypropylene/Ethylene-Propylene Block Polymers", Polym. Eng. Sci., 24, 1319, 1984.

Dumoulin, M.M., Carreau, P.J., and Utracki, L.A., "Rheological Properties of Linear Low Density Polyethylene/Polypropylene Blends - Part 2: Solid State Behavior", Polym. Eng. Sci., 27, 1627, 1987.

Einstein, A., "Eine neue Bestimmung der Molekuldimension", Ann. Physik, 19, 289 (1906) and 34, 591, 1911.

Eveson, G.F., "Rheology of Dispersed Systems", C.C. Mills, Ed., Academic Press, New York, 1959.

Ferry, J.D., "Viscoelastic Properties of Polymers" , 3rd Edition, Wiley, New York, 1980.

Faulkner, D.L., and Schmidt, L.R., "Glass Bead-Filled Polypropylene Part 1: Rheological and Mechanical Properties", Polym. Eng. Sci., 17, 657, 1977.

Gupta, R.K., and Seshadri, S.G., "Maximum Loading Levels in Filled Liquid Systems", J. Rheol., 30, 503, 1986.

Hamielec, L.A., "Rheological Properties of Suspensions of Newtonian and Non-Newtonian Fluids", Polym. Eng. Sci., 26, 111, 1986.

Kamal, M.R. and Mutel, A.T., "Rheological Properties of Suspensions of Newtonian and Non-Newtonian Fluids", J. Polym. Eng., 5, 293, 1986.

Kitano, T., Kataoka, T. and Shirota, T., "An Empirical Equation of the Relative Viscosity of Polymer Melts Filled with Various Inorganic Fillers", Rheol. Acta, 20, 207, 1981.

Lakdawala, K. and Salovey, R., "Rheology of Polymers Containing Carbon Black", Polym. Eng. Sci., 27, 1035, 1987.

Lara A.J., and Schreiber, H.P., "Specific Interactions and Adsorption of Film-Forming Polymers", submitted to J. Coat. Tech., 1990.

Malik, T.M., Carreau, P.J., Grmela, M., and Dufresne, A., "Mechanical and Rheological Properties of Carbon Filled Polyethylene", Polymer Composites, 9, 412, 1988.

Malik, T.M., and Prud'homme, R.E., "Solid-State Relaxation Properties of Poly(α-methy-α-η-propyl-β-propyolactone)/Poly(vinyl Chloride) Miscible Blends", Macromolecules, 16, 311, 1983.

Maron, S.H., and Pierce, P.E., "Application of Ree-Eyring Generalized Flow Theory to Suspensions of Spherical Particles", J. Colloid Sci., 11, 80, 1956.

Metzner, A.B., "Rheology of Suspensions in Polymeric Liquids", J. Rheol. 29, 739, 1985.

Menendez, H., and White, J.L., "A Wide-Angle X-Ray Diffraction Method of Determining Chopped Fiber Orientation in Composites with Application to Extrusion Through Dies", Polym. Eng. Sci., 24, 1051, 1984.

Minagawa, N., and White, J.L., "Coextrusion of Unfilled and TiO_2-filled Polyethylene: Influence of Viscosity and Die Cross-Section on Interface Shape", Polym. Eng. Sci, 15, 825, 1975.

Nielsen, L.E., "Polymer Rheology", Marcel Dekker, New York, 1977.

Nielsen, L.E., "Mechanical Properties of Polymers and Composites", Marcel Dekker , New York, 1974.

Nguyen, H.X. and Ishida, H., "Poly(Aryl-Ether-Ether-Ketone) and Its Advanced Composites: A Review", Polym. Compos., 8, 57, 1987.

Otsubo, Y., "Effect of Polymer Adsorption on the Rheological Behavior of Silica Suspensions", J. Colloid and Interf. Sci., 112, 380, 1986.

Poslinski, A.J., Ryan, M.E., Gupta, R.K., Seshadri, S.G., and Frechette, F.J., "Rheological Behavior of Filled Polymeric Systems", J. Rheol., 32, 703 and 751, 1988.

Romanini, D., "Synthesis Technology, Molecular Structure and Rheological Behavior of Polyethylene", Polym. Plast. Technol. Eng., 19, 201, 1982.

Speed, C.S., "Formulating Blends of LLDPE and LDPE to Design Better Film", Plast. Eng., P. 39, July 1982.

Strobel, W., Kunstsoffe, 77, 967, 1987.

Thomas, D.G., "Transport Characteristics of Suspension: Note on the Viscosity of Newtonian Suspensions of Uniform Spherical Particles", J. Colloid Sci., 20, 267, 1965.

Utracki, L.A., in "Rheological Measurement", edited by Collyer, A.A., and Clegg, D.W., Elsevier Applied Science, London and New York, 1988.

Yeh, P.L. and Birley, A.W., "Blends of Polypropylene and Linear Low Density Polyethylene", Plast. Rubb. Proc. Appl., 5, 249, 1985.

Chapter 9

Simulation of Particle Deposition in Fluid Flow

B. V. RAMARAO
Department of Paper Science and Engineering
SUNY—College of Environmental Science and Forestry
Syracuse, New York 13210, USA

C. TIEN
Department of Chemical Engineering and Materials Science
Syracuse University
Syracuse, New York 13244-1240, USA

1. Introduction

The purpose of this chapter is to present a general
treatment of the deposition of diffusive aerosols from a
fluid flowing past a collector. Our focus is specifically
on a quantitative evaluation of the effect of the various
parameters and variables on the deposition phenomenon. By
diffusive aerosols, we refer to those aerosols which
exhibit Brownian motion in a significant but not
necessarily dominant way. This condition is usually met
for relatively large sub-micrometer particles in gas flow
at relatively high velocity and at moderate temperatures
and pressures.

There are several practical reasons for studying the
deposition of diffusive aerosols in fluid flow. First, it
occurs in a number of important natural processes; for
instance, the deposition of toxic particles and droplets
from the atmosphere to the earth. Secondly, the
phenomenon may be used as a model depicting certain man-
made processes. For instance, filtration of aerosols in
granular or fibrous beds can be idealized by considering
a filter bed as an assembly of collectors of appropriate
geometry. The retention of particles by filters may then
be considered as resulting from suspensions flowing past
these collectors. The performance of a filter can be
directly related to the deposition rates within it.

The dynamics of particles in a flowing gas medium may be
characterized by their positions, velocities and concentrations which, in principle, may be obtained if the
particle trajectories are known. A quantity of practical
interest and importance in many cases is the particle
deposition flux. Estimation of deposition flux can be
made from a knowledge of particle trajectories in the
vicinity of the collector surface together with the
information concerning particle collector interactions.
The use of the so-called trajectory analysis concept in
estimating the collection efficiencies in granular and
fibrous filtration is an example of this concept.

Particle trajectories can in general be found if the types of forces acting on the particle and their magnitudes are known.

Particle trajectories may be determined from the solution of the equations of motion formulated based on Newton's law. For aerosols of supra-micrometer size, Brownian motion is relatively unimportant and solution of the equations of particle motion is straightforward. On the other hand, for diffusive aerosols which experience significant Brownian motion, trajectory determination becomes more complicated because of the stochastic nature of Brownian motion. Generally speaking, particle trajectory is commonly conceived as a deterministic concept. Given a set of conditions for a particle (namely, the forces acting on the particle and the initial conditions of the particle), its trajectory is unique. When one of the forces acting on the particle is stochastic, this deterministic concept, of course, becomes invalid and a different concept must be applied. As an example, consider the problem of estimating particle deposition flux of diffusive aerosols when the Brownian motion is dominant. It is customary to consider these aerosol particles as constituting a molecular species in the gas phase. The deposition problem becomes a mass transfer problem wherein the deposition flux is the equivalent of the mass flux. The flux can be predicted from the solution of the convective diffusion equation (Friedlander, 1977). This method can be further extended to include the effect due to the presence of other forces if these forces may be represented as the gradients of their respective force potentials.

For the more general case where the effect of particle inertia must be considered in addition to Brownian diffusion and other forces, analysis of the particle behavior, in general, and the estimation of the deposition flux, in particular, become more difficult. Three different methods for studying this kind of problem have been proposed in recent years. Yuu and Jotaki (1976) developed an approximate method for examining the combined effect of inertial impaction and Brownian diffusion on particle deposition in an ideal stagnation point flow. In their analysis, Yuu and Jotaki used the conventional convective diffusion equation as suggested by Friedlander but replaced the fluid velocity with the particle velocity which, in turn, was estimated from particle trajectory determinations assuming that the Brownian motion effect was negligible. The inherent assumption in their approach is that the particle trajectories replace the fluid streamlines. The solution of the resultant hybrid convective diffusion equation gives the expressions for the deposition flux and provides estimates of the collection efficiency.

The second approach which may be termed as the pseudo-continuum approach was used by Fernandez de la Mora and Rosner (1983,1984) for their studies of deposition to a planar surface in a stagnation flow and to a spherical collector in Stokes flow. The pseudo-continuum approach considers a fluid-solid suspension to be composed of two continua: a gas or fluid phase for which the conventional equations of continuum mechanics hold and a hypothetical particle phase which is characterized by a second set of flow equations. In formulations such as this, a closure problem usually arises. In this case, it becomes necessary to assume constitutive equations relating the pressure and stress tensor of the particle phase with the relevant kinematic quantities. The formulations of these expressions can only be done arbitrarily.

Both of the two methods mentioned above have their limitations. More recently, Gupta and Peters (1985,1986) considered the deposition of particles from suspensions flowing past a spherical collector by solving the relevant Langevin equations using a simulation approach based on the Brownian dynamics. A major potential advantage of their method is that it is rigorous and is conceptually simple. Although the simulation method of Gupta and Peters' requires significant computational effort, more recent studies by the present authors indicate that it is possible to reduce the computations. Because of these considerations, the solution of the Langevin equations using the simulation approach based on the Brownian dynamics will form the basis of the present analysis.

The principal focus of the presentation given below is placed on the determination of particle trajectories and deposition fluxes of diffusive aerosols using the Brownian dynamics simulation method. First, equations describing the motion of aerosol particles in Brownian motion will be derived and their solutions will be presented. Next, the procedure which may be used in converting the equation of particle motion into an equation describing the particle probability field (or concentration field) will be discussed. With these discussions as background, the problem of deposition of diffusive aerosols will be discussed in general and the methods used for its solution under various conditions will be developed.

2. DYNAMICS OF DIFFUSIVE PARTICLES

In studying particle dynamics, the main interests are the trajectories of the particles and their velocities from which one can readily determine the deposition fluxes and particle concentrations. When one uses the convective-diffusion equation to determine particle concentrations and fluxes, the implicit assumption involved is that particle trajectories and fluid streamlines are the same (i.e. no inertial effects). When particles experience significant inertial forces, particle velocity may become quite different from that of the fluid. In such cases, the convective diffusion approach fails and one has to investigate the problem from a different viewpoint.

2.1 Langevin Equation: Lagrangian View

The Lagrangian view of particle motion in principle, can be extended to include the diffusive motion of the Brownian particles thus providing a convenient framework for examining the combined effects of particle inertia and Brownian motion on particle deposition. In the Lagrangian approach, one is concerned with the trajectory traveled by a single particle subject to given initial conditions in the flow field in isolation of all other particles. The motion of the particle is dictated by the laws of classical dynamics i.e. force and torque balances. The torque balance may be ignored for small spherical particles since their rotation is generally insignificant.

A force balance for a particle may be written as:

$$m \frac{dv}{dt} = F_d + F_e + F_r \qquad (1)$$

where the particle inertia given on the left hand side is balanced against the sum of three different 'kinds' of forces, F_d, F_e, and F_r denote respectively the particle mass and velocity. Regarding the three forces, F_d is the total hydrodynamic drag exerted by the fluid on the particle while F_e represents any external force field acting on the particles. If F_e is conservative, as is usually the case, it can be expressed as the gradient of a potential. Familiar examples are the gravitational attraction, buoyancy and Coulombic attraction type of electrostatic forces. When the forces acting on the particle include only F_d & F_e or Equation (1) is deterministic in nature and is a first order ODE in the particle velocity v or a second order ODE if the position vector r of the particle can be considered as the dependent variable instead of v. Equation (1) can be integrated numerically with a given set of initial conditions and the particle trajectory r and velocity v can be found.

The third force F_r, is a random force exerted on the particle. This force is taken to represent the combined effect of a large number of fluid molecular collisions experienced by the particle, which cause the random Brownian motion. When F_r is present and is a significant part of the total force, Equation (1) becomes a stochastic differential equation which commonly is referred to as the Langevin equation (See e. g. Chandrasekhar, 1943). Under the condition that F_r is significant, Equation (1) represents a random function whose probability distribution can be obtained from its solution. Similarly, the particle trajectory is also stochastic in nature. One useful concept which will help in later analysis is to view F_r as a non-unique forcing function (one sample out of an ensemble of such functions). This implies that r [and equivalently, v] is also non-unique.

Integration of Equation (1) leads to a probability density function giving the probability, $W(r)$, of a given particle occupying a location r in space. For time being we will be concerned primarily with solutions of the Langevin equation in a stochastic sense - specifying the probability density distribution for a particle occupying a given position in space with a given velocity. Insofar as Brownian dynamics simulations are concerned, the important distributions are: $W(r,t;r_0,t_0)$, the probability density distribution of a particle occupying a position r, at a time t, with an initial position r_0, at time t_0; $W(v,t;v_0,t_0)$, the corresponding probability density for the particle velocity and, $W(r,v,t;r_0,v_0,t)$ the joint distribution of r and v simultaneously under the specified initial conditions. Under the condition that particle inertia and Brownian diffusion are both important, the use of the joint probability is necessary in determining the trajectory. In this case, the velocity of the particle is no longer decoupled from the particle position. The random force can be written as the product of the particle mass and a random Brownian acceleration $A(t)$. By doing so the random nature of F_r (t) is now ascribed to $A(t)$. We assume that $A(t)$ is independent of $v(t)$. Furthermore $A(t)$ fluctuates rapidly as compared to the variations in v (see Chandrasekhar). The random acceleration $A(t)$ has a mean of zero and its auto-correlation is an impulse function. Thus,

$$<A(t)> = 0 \tag{2}$$

$$<A\ (t)\ A\ (t+\tau)> = K\ \delta\ (t - \tau) \tag{3}$$

Equation (3) implies that the magnitude of the Fourier transform of $A(t)$ is a constant. Thus, all frequencies (from $-\infty$ to ∞) are represented equally in a trace of $A(t)$. Such a function is termed white noise indicating the equal strengths of all frequencies. $A(t)$ therefore represents a Gaussian white noise process in stochastic terms. In the following, we present solutions of Equation (1) under certain specific conditions.

Case 2.1.1 Stationary fluid without external forces

If we assume that the Brownian particle is moving in a
stationary fluid and further that the drag force acting
on this particle is given by Stokes' law (at low particle
Reynolds number), Equation (1) becomes

$$m \frac{d\underline{v}}{dt} = - m\beta\underline{v} + m \underline{A} (t) \tag{4}$$

where β is the friction coefficient.

With the equation of particle motion written in the form
of Equation (4), we are conceptualizing Brownian motion
as consisting of two time scales. One is a fast scale, of
the order of the time between successive molecular
collisions. In liquids this time scale is of the order of
10^{-19} s (Lin, 1989). The second is a slower time scale
manifesting the adjustment of fluid velocity responding
to changes in its momentum. The order of this time scale
is given by $1/\beta$. The two time scales do not overlap.

2.1.1.1 Solution for particles without inertia

When particle inertia is ignored, Equation (4) is
reduced to

$$\underline{v} = + \frac{1}{\beta} \underline{A} (t) \tag{5}$$

and the distribution of $\underline{v}$ is the same as that of $\underline{A}$.
Further, from Equation (5),

$$\underline{r} (t) = + \frac{1}{\beta} \int_{o}^{t} \underline{A} (\zeta) d\zeta \tag{6}$$

which identifies $\underline{r}(t)$ as an integral of the Brownian
process $\underline{A}(\zeta)$. This is called a Weiner process (Papoulis,
1986). The distribution of $\underline{r}$ is Gaussian with mean zero.
The variance of $\underline{r}(t)$ can be determined by applying
Theorem (shown in Appendix A) for $\underline{A}(t)$. The distribution
of $\underline{r}$ is:

$$W(\underline{r}) = \frac{1}{(4\pi q \frac{t}{\beta^2})^{3/2}} \exp \left(- \frac{\underline{r} \cdot \underline{r}}{4q \frac{t}{\beta^2}}\right) \qquad (7)$$

where q is given by the ratio $\beta kT/m$.

2.1.1.2 Solution for inertial particles

When particle inertia is significant, we can apply Equation (4) to the particles directly. Integration of this equation with repsect to time gives us the velocity of the Brownian particle at time t, subject to the condition that at time t=0, it has a velocity $\underline{v}_0$ as

$$\underline{V} = \underline{V}_0 e^{-\beta t} + \underline{R}_1(t) \qquad (8)$$

where, we have expressed the particular integral

$$e^{-\beta t} \int_0^t \underline{A}(\zeta) d\zeta e^{\beta \zeta} = \underline{R}_1(t) \qquad (9)$$

by the random vector $\underline{R}_1(t)$. We may consider $\underline{R}_1$ to represent a random increment to the particle's initial velocity $\underline{v}_0$. The probability distribution of $\underline{R}_1$ can be given by

$$W(\underline{R}_1, t; \underline{v}_0) = \left[\frac{m}{2\pi kT(1-e^{-2\beta t})}\right]^{3/2} \exp \left[\frac{-m \underline{R}_1 \cdot \underline{R}_1}{2kT(1-e^{-2\beta t})}\right] \qquad (10)$$

We may integrate Equation (8) with the understanding that $\underline{r} = d\underline{v}/dt$, further to obtain

$$\underline{r} = \underline{r}_0 + \frac{\underline{v}_0}{\beta} (1-e^{-\beta t}) + \underline{R}_2(t) \qquad (11)$$

where r_0 is the particle position vector at the initial time t=0 and $R_2(t)$ represents the particular integral

$$\int_0^t R_1(\zeta)\, d\zeta = R_2(t) \tag{12}$$

The distribution of this integral can be obtained by expansion by parts and application of the theorems in appendix A. The result is given as

$$W[R_2, t; v_0, r_0] = [\frac{m\beta^2}{2\pi kT}(2\beta t - 3 + 4e^{-\beta t} - e^{-2\beta t})]^{3/2} \tag{13}$$

$$\exp[-\frac{m\beta^2}{2kT}\frac{R_2 \cdot R_2}{(2\beta t - 3 + 4e^{-\beta t} - e^{-2\beta t})}]$$

Case 2.1.2 Fluid in motion with external forces

The Langevin equation for this case may be written as

$$\frac{dv}{dt} = \beta(u - v) + \frac{F_e}{m} + A(t) \tag{14}$$

The drag force on the particle is assumed to be modeled by Stokes' law and is thus proportional to the relative velocity between the fluid and the particles. This equation is difficult to solve for any general field. However, for the particular case when u and F_e are constant or linear in particle position, the equation can be solved.

2.1.2.1 Constant Fluid Velocity for Inertialess Particles

The equation for the particle velocity under conditions when inertia can be neglected is obtained from Equation (14). If the external force is absent, one has

$$v = u + \frac{1}{\beta} A(t) \tag{15}$$

and the particle position is obtained by substituting dr/dt for v in the above equation.

$$\frac{d\underline{r}}{dt} = \underline{u} + \frac{1}{\beta} \underline{A} (t) \qquad (16)$$

Integrating this equation subject to the condition that at t=0, $\underline{r} = \underline{r}_0$, we obtain

$$\underline{r} = \underline{r}_o + \underline{u}t + \frac{1}{\beta} \underline{R} (t) \qquad (17)$$

where $\underline{R} = \int_0^t \underline{A} (\zeta) d\zeta$. The distribution of $\underline{R}(t)$ is given by an equation similar to (7)

$$W (\underline{R};\underline{r}_o) = \frac{1}{(4\pi qt)^{3/2}} \exp (-\frac{\underline{R}.\underline{R}}{4qt}) \qquad (18)$$

2.1.2.2 Constant Fluid Velocity for Inertial Particles

With particle inertia included, and also supposing $\underline{u}$ to be a constant, the complete Langevin equation in the form of Equation (14) can be solved. The solution is described by Peters & Gupta (1985) and is given for the particle velocity as

$$\underline{v} = \underline{v}_o e^{-\beta t} + \frac{\underline{u}}{m} (1-e^{-\beta t}) + \frac{\underline{F}_e}{m\beta} (1-e^{-\beta t}) + \underline{R}_1 (t) \qquad (19)$$

where

$$\underline{R}_1(t) = \int_o^t e^{\beta(\zeta-t)} \underline{A} (\zeta) d\zeta \qquad (20)$$

Substituting $d\underline{r}/dt$ for $\underline{v}$ in Equation (19) and solving, we get

$$\underline{r} = \underline{r}_o + \frac{v_o}{\beta} (1-e^{-\beta t}) + \frac{u}{m} [t - \frac{1}{\beta} (1-e^{-\beta t})]$$

$$+ \frac{F_e}{m\beta} [t - \frac{1}{\beta} (1-e^{-\beta t})] + \underline{R}_2 (t) \qquad (21)$$

where

$$\underline{R}_2 (t) = \int_o^t [\int_o^n e^{\beta \zeta} \underline{A}(\zeta) \, d\zeta] e^{-\beta n} \, dn \qquad (22)$$

The integrals $\underline{R}_1$ and $\underline{R}_2$ are stochastic in nature. It is possible for us to determine their joint distribution as shown by Peters & Gupta (1984) as

$$W(\underline{R}_1, \underline{R}_2) = \frac{1}{8\pi^3 (FG-H^2)^{3/2}} \exp [- \frac{G \underline{R}_1 \cdot \underline{R}_1 - 2H \underline{R}_1 \cdot \underline{R}_2 + F \underline{R}_2 \cdot \underline{R}_2}{2(FG-H^2)}] \qquad (23)$$

where

$$F = \frac{3kT}{m\beta^2} [2\beta t - 3 + 4e^{-\beta t} - e^{-2\beta t}] \qquad (24)$$

$$G = \frac{3kT}{m} [1 - e^{-2\beta t}] \qquad (25)$$

$$H = \frac{3kT}{m} [1 - e^{-\beta t}]^2 \qquad (26)$$

2.1.2.3 Linear Fluid Velocity

For an ideal stagnation point flow the fluid flow field is given as

$$u_x = \alpha x \qquad (27a)$$

$$\underline{u} =$$

$$u_y = -\alpha y \qquad (27b)$$

where α is a flow constant. Denoting i to be the x or y component ($\delta = 1$ for x and ($\delta = 1$ for y and $\delta = -1$ for y in above), $\underline{u}$ can be written as $\delta \alpha\, x_i$. Equation (4) becomes

$$\frac{dv_i}{dt} = \beta(\delta \alpha x_i - v_i) + \frac{F_i}{m} + A_i\,(t) \qquad (28)$$

Note further that $dx_i/dt = v_i$, one has

$$\frac{d^2 x_i}{dt^2} + \beta\,\frac{dx_i}{dt} - \delta\beta\alpha\,x_i = \frac{F_i}{m} + A_i\,(t) \qquad (29)$$

The solution of this equation obtained by Ramarao & Tien (1990b) is

$$x_i = \frac{1}{\beta_{1i}}\,[v_{io} - x_{io}\,(-\frac{\overline{\beta_{1i}+\beta}}{2}) - \frac{2\,f_i}{(\beta_{1i}-\beta)}]e^{m_{1i}t}$$

$$+ \frac{1}{\beta_{1i}}\,[-\{v_{io} - x_{io}\,\frac{(\beta_{1i}-\beta)}{2} + \frac{2\,f_i}{(\beta_{1i}+\beta)}\}\,]e^{m_{2i}t} - \frac{4\,f_i}{(\beta^2+\beta_{1i}^2)} + x_i^R$$

$$\qquad (30)$$

$$m_{1i} = (\beta_{1i} - \beta)/2 \tag{31}$$

$$m_{2i} = -(\beta_{1i} + \beta)/2 \tag{32}$$

$$\beta_{1i}^2 = \beta^2 + 4\delta\alpha\beta \tag{33}$$

$\underline{v}$ may be determined by differentiating Equation (30) with respect to time.

$$v_i = \frac{m_{1i}}{\beta_{1i}} [v_{io} - x_{io} (-\frac{\overline{\beta_{1i} + \beta}}{2}) - \frac{2 f_i}{(\beta_{1i} - \beta)}] e^{m_{1i}t}$$

$$- \frac{m_{2i}}{\beta_{1i}} [v_{io} - x_{io} \frac{(\beta_{1i} - \beta)}{2} + \frac{2 f_i}{(\beta_{1i} + \beta)}] e^{m_{2i}t} \tag{34}$$

The distribution $W(\underline{r}, v; \underline{r}_0, \underline{v}_0, \underline{t}_0)$ can be determined from Equation (23). This distribution is bivariate Gaussian with a mean of zero. The equation for this distribution is given by Equation (23) with $\underline{r}$ and $\underline{v}$ replacing $\underline{R}_1$ and $\underline{R}_2$. The covariance matrices of each component (x^R, v_x^R) and (x^R, v_y^R) then completely determine the distribution.

$$\text{cov} (x_i^R, v_i^R) = [\begin{matrix} F_i & H_i \\ H_i & G_i \end{matrix}] \tag{35}$$

$$F_i = \frac{2q}{\beta_{1i}^2} \{ \frac{1}{(\beta_{1i} - \beta)} e^{(\beta_{1i} - \beta)t} - \frac{1}{(\beta_{1i} + \beta)} e^{-(\beta_{1i} + \beta)t} + \frac{2}{\beta} (e^{-\beta t} - 1) + 2\beta/(\beta^2 - \beta_{1i}^2) \} \tag{36}$$

$$G_i = \frac{2q}{\beta_{1i}^2} \{ \frac{(\beta_{1i} - \beta)}{4} e^{(\beta_{1i} - \beta)t} - \frac{(\beta_{1i} + \beta)}{4} e^{-(\beta_{1i} + \beta)t} + \frac{(\beta^2 - \beta_{1i}^2)}{\beta} (e^{-\beta t} - 1) + \beta/2 \} \tag{37}$$

$$H_i = \frac{1}{2\beta_{1i}^2} \{ e^{(\beta_{1i} - \beta)t/2} - e^{-(\beta_{1i} + \beta)t/2} \}^2 \tag{38}$$

In the above equations, (30)-(38), i=x or y. Further we have denoted $f_i = \underline{F}_e / m$. This completes the solution of (28).

Case 2.1.3   Basset-Boussinesq-Oseen Equation

The drag force $\underline{F}_d$ of the equation of particle motion [i.e. Equation (1)] is commonly assumed to be given by the Stokes law even for accelerating particles. Strictly speaking, this practice is not valid since Stokes law gives the drag force acting on a stationary particle suspended in a flow which is uniform far away from the particle. When a particle is placed in an arbitrary flow field and in the absence of external boundaries, the net hydrodynamic drag exerted on the particle is the sum of Stokes' drag and another term (called Faxen's force) that is proportional to the Laplacian of the velocity field evaluated at the center of the particle. For the case of an accelerating particle, the net drag force further includes a virtual mass term and a Basset force term. The complete equation of particle motion has been referred to as the Basset-Boussinesq-Oseen equation [Hinze(1975), Corrsin & Lumley (1956), Tchen (1947)].

$$m \frac{dv}{dt} = m \left(\frac{du}{dt} - \nu \nabla^2 \underline{u}\right) - \frac{m}{2}\left(\frac{dv}{dt} - \frac{u}{t} - \underline{v}.\nabla\underline{u}\right) + \underline{K}\ (\underline{r})$$

$$- 6\pi\mu r_p(\underline{v} - \underline{u}) - \frac{6\pi\mu r_p^2}{\sqrt{\pi\nu}} \int_{-\infty}^{t} \left(\frac{dv}{d\tau} - \frac{\partial u}{\partial\tau} - \underline{v}.\nabla\underline{u}\right)/ \sqrt{t-\tau}\ d\tau \qquad (39)$$

In this equation, the particle inertial force appears on the left hand side. On the right hand side, each of the terms corresponds in succession to (a) the Faxen's law drag term, (b) the additional drag due to the virtual mass of the particle, (c) an external force field $\underline{K}(\underline{r})$, (d) the Stokes' hydrodynamic drag and (e) the Basset force (integral terms) respectively. For aerosol particles, the extra terms resulting from the use of the correct drag force expression are relatively unimportant in most instances. However, for particles suspended in liquids, these extra terms could be of considerable magnitude and significantly affect particle deposition and particle trajectories. As Equation (39) is an integro-differential equation of second order, obtaining its solution in general is difficult. Under certain special situations it can be solved analytically using integral transform methods as shown below.

2.1.3.1 Particle coming to rest in a stationary fluid

Consider a particle that moves with a uniform velocity initially v_0 and comes to rest in a stationary fluid. An example of this situation is a particle which is moving at a constant terminal velocity under the influence of an external force. The force is removed at a certain instant and the particle is allowed to come to rest. The Basset-Boussinesq-Oseen equation becomes

$$3/2m \frac{dv}{dt} + 6\pi r_p v - K(r) - \frac{6\pi\mu r_p^2}{\sqrt{\pi\nu}} \int_{-\infty}^{t} \frac{dv/d\tau}{\sqrt{t-\tau}} d\tau = 0 \qquad (40)$$

The solution of this equation (with K set to zero) obtained by Ramarao & Tien (1990a) is

$$\frac{v}{v_0} = \frac{1}{(\mu_1 - \mu_2)} \{ [a_0 + \frac{a_2}{\mu_2}] \sqrt{\mu_2} \, e^{\mu_2 t} \, \text{erfc} \, (\sqrt{\mu_2 t}) - [a_0 + \frac{a_2}{\mu_1}] \sqrt{\mu_1} \, e^{\mu_1 t} \, \text{erfc}(\sqrt{\mu_1 t}) \qquad (41)$$

where a_0, a_1, a_2 are the constants

$$a_0 = 3 \, m/2 \qquad (42)$$

19

$$a_1 = 6\pi\mu r_p \qquad (43)$$

$$a_2 = 6\pi\mu r_p^2 / \sqrt{\pi\gamma} \qquad (44)$$

and μ_1, μ_2 are given as the roots of the polynomial

$$(sa_o + \sqrt{s}a_2 + a_1) = (\sqrt{s} - \mu_1)(\sqrt{s} - \mu_2) \qquad (45)$$

The equation for particle position can be obtained by integration of the velocity.

2.1.3.2 Particle in Linear Flow

When a particle is in a linear flow (e.g. ideal stagnation point flow), the Basset-Boussinesq-Oseen equation remains linear as in the case of stationary fluid. The equation is

$$3/2m \frac{d^2x_i}{dt^2} + (6\pi\mu r_p + \frac{\delta\alpha m}{2})\frac{dx_i}{dt} + \delta\alpha x_i \, 6\pi\mu r_p$$

$$+ \frac{6\pi\mu r_p^2}{\sqrt{\pi\nu}} \int_{-\infty}^{t} \frac{d^2x/d\tau^2 - \delta\alpha \, dx_i/d\tau}{\sqrt{t-\tau}} \, d\tau = 0 \qquad (46)$$

The solution of Equation (46) was also obtained by Ramarao & Tien (1990a) and is given as

$$x_i(t) = \sum_{i=1}^{4} \frac{(b_3\mu_i^2 + b_2\mu_i + b_1|\mu_i + b_o)}{\prod_{\substack{\ell \neq j \\ j=1}}^{4} (\mu_j - \mu_\ell)} \{\frac{1}{\sqrt{\pi t}} - \mu_i e^{\mu_i^2 t} erfc(\mu_i\sqrt{t})\} + \frac{b_1}{\mu_1\mu_2\mu_3\mu_4\sqrt{\pi t}} \qquad (47)$$

$$(a_o s^2 + a_2 s^{3/2} + a_1 s - a_4\sqrt{s} + a_3) = (\sqrt{s} + \mu_1)(\sqrt{s} + \mu_2)(\sqrt{s} + \mu_3)(\sqrt{s} + \mu_4) \qquad (48)$$

where

$$a_o = 3/2 \; m \tag{48a}$$

$$a_1 = 6\pi\mu r_p + \delta\alpha \, m/2 \tag{48b}$$

$$a_2 = 6\pi\mu r_p^{\,2}/\sqrt{\pi\nu} \tag{48c}$$

$$a_3 = 6\pi\mu r_p \, \delta\alpha \tag{48d}$$

$$a_4 = \delta\alpha \; a_2 \tag{48e}$$

$$b_o = a_o v_o + a_1 x_o \tag{48f}$$

$$b_1 = a_2 v_o - a_4 x_o \tag{48g}$$

$$b_2 = a_2 x_o \tag{48h}$$

$$b_3 = a_o x_o \tag{48i}$$

Case 2.1.4 Basset-Langevin Equation

In most cases, the additional unsteady components of the drag force may be ignored. However, certain pathological cases exist when both the unsteady drag and the Brownian motion term need to be considered. Such situations can be investigated using the equation of particle motion in the form of Equation (39) with a Brownian force appended to it. The resulting equation has been termed the Basset-Langevin equation (Arminski & Weinbaum, 1978). It is a stochastic integro-differential equation and is given by Ramarao & Tien (1990c) as

$$m\frac{d\underline{v}}{dt} - \frac{m}{2}\left(\frac{d\underline{u}}{dt} - \nu\nabla^2\underline{u}\right) + \frac{m}{2}\left(\frac{d\underline{v}}{dt} - \frac{\partial\underline{u}}{\partial t} - \underline{v}\cdot\nabla\underline{u}\right) - \underline{K}(\underline{r})$$

$$+ \; 6\pi\mu r_p(\underline{v}-\underline{u}) + \frac{6\pi\mu r_p^{\,2}}{\sqrt{\pi\nu}} \int_{-\infty}^{t} \left(\frac{d\underline{v}}{d\tau} - \frac{\partial\underline{u}}{\partial\tau} - \underline{v}\cdot\nabla\underline{u}\right)/\sqrt{t-\tau}\; d\tau = m\,\underline{A}\,(t) \tag{49}$$

2.1.4.1 Particle coming to rest in stationary fluid

For the case of stationary fluid (setting K = 0), Equation (49) becomes

$$3/2 \ m \ \frac{dv}{dt} + 6\pi\mu r_p \underline{v} - \frac{6\pi\mu r_p^2}{\sqrt{\pi}\nu} \int_{-\infty}^{t} \frac{d\underline{v}/d\tau}{\sqrt{t-\tau}} \ d\tau = m \ \underline{A} \ (t) \tag{50}$$

The solution of Equation (50) may be written as $\underline{v} = \underline{v}^D + \underline{v}^R m$ where $\underline{v}^D$ is given by Equation (41) and the distribution of $\underline{v}^R$ is

$$W\{\underline{v}^R\} = \frac{1}{[2\pi q \int_o^t \phi^2(\zeta)d\zeta]^{3/2}} \ \exp \ \{- \frac{\underline{v}^R \cdot \underline{v}^R}{2q \int_o^t \phi^2(\zeta)d\zeta} \}$$

where

$$\phi(\zeta) = \frac{1}{(\mu_1 - \mu_2)} \{\sqrt{\mu_1} \ e^{\mu_1 t} \ erfc \ (\sqrt{\mu_1}\sqrt{t}) - \sqrt{\mu_2} \ e^{\mu_2 t} \ erfc \ (\sqrt{\mu_2}\sqrt{t})\} \tag{51}$$

The constants μ_1, μ_2 are given as in Equations (41-45) again.

2.1.4.2 Particle in a Linear Flow

For the case of ideal stagnation point flow, [i.e. Equation (27)], Equation (49) becomes

$$3/2 \ m \ \frac{d^2x_i}{dt^2} + (6\pi\mu r_p + \frac{\delta\alpha m}{2}) \ \frac{dx_i}{dt} + \delta\alpha x_i \ 6\pi r_p + \frac{6\pi r_p^2}{\sqrt{\pi}\nu} \int_{-\infty}^{t} \frac{d^2x_i/d\tau^2 - \delta\alpha dx_i/d\tau}{\sqrt{t-\tau}} \ d\tau$$

$$= m \ A_i \ (t) \tag{52}$$

207

Its solution is given similar to the previous case as $\underline{x} = \underline{x}^D + \underline{x}^R$ $\underline{x}^D$ is given by Equation (47) and the distribution of $\underline{x}^R$ by

$$W\{\underline{x}^R\} = \frac{1}{[2\pi q \int_0^t \phi^2(t-\zeta)d\zeta]^{3/2}} \exp\{-\frac{\underline{x}^R \underline{x}^R}{2\,q \int_0^t \phi^2(t-\zeta)d\zeta}\} \quad (53)$$

$$\phi(\zeta) = -\sum_{i=1}^{4} \frac{1}{\prod_{j\neq i}^{4}(\mu_j-\mu_i)} \mu_i\, e^{\mu_i^2 t}\, \text{erfc}\,(\mu_i\sqrt{t}) \quad (54)$$

The constants are given by Equations (48, 48a-i).

It must be remarked here that although the above equation has been used as a model for studying the effects of unsteady drag forces on a Brownian particle, there is considerable controversy regarding its validity. The Basset memory term is a frequency dependent friction force. When a particle experiences an acceleration, the drag force exerted by the fluid around it is a function not only of the instantaneous velocity field but also the history of the changes in such a velocity field. The nature of the Brownian motion is generally assumed to be Markovian i.e. that the small random force experienced by the particle due to molecular collisions is assumed to be a white noise process. This means that this Brownian collision force has no memory and thus allows us to assume that the path of the Brownian particle is dependent upon its immediate past position only and not upon its history. The inclusion of the memory integral in the force balance is in contradiction to this assumption. In order to be consistent we then have to assume that the Brownian motion is no longer Markovian but has a long memory (see for e.g. Hinch, 1975). Thus, the above form of the Basset - Langevin equation is a gross approximation at best.

2.2 Fokker-Planck Equation: Eulerian View

For diffusive aerosols, it is possible to approach the problem of determining particle probability densities from an Eulerian field perspective as opposed to the Lagrangian view considered by the Langevin equation as discussed above. The key element here is to obtain a

partial differential equation for the particle
probability density directly from a knowledge of the
probability density of occurrence of chosen velocity
(also called the transition probability).

As an example, we derive the formal Fokker-Planck
equation for the case of Brownian particles with
significant inertia placed in an arbitrary flow field.
The development detailed below follows the general
development given by Chandrasekhar (1943) on the Fokker-
Planck equation for a Brownian harmonic oscillator. We
recognize that in the dynamic equation for an oscillator,
if the restoring force is replaced by the term $\beta \underline{u}$, the
dynamic equation for an inertial aerosol particle placed
in a fluid flow is obtained. Let us consider the Brownian
motion of an ensemble of such particles. We will assume
that the particle concentration is low such that the
particles do not interact with each other. Let us focus
our attention on the motion during a time interval within
which a very large number of particle displacements
occur. However, Δt is sufficiently small such that the
velocities of the Brownian particles themselves do not
change appreciably. Further, the fluid velocity may also
be assumed to be a constant over the interval Δt. Under
these conditions, we can expect the distribution of the
particle velocity $\underline{v}$ at time $t+\Delta t$ to be derivable from the
same probability at t and also a knowledge of the like-
liness of occurrence of all possible increments to the
velocity at the earlier time t. This is expressed
formally as

$$W(\underline{r},\underline{v},t+\Delta t) = \int \int W\ (\underline{r}-\Delta\underline{r},\underline{v}-\Delta\underline{v},t)\ \psi(\underline{r}-\Delta\underline{r},\underline{v}-\Delta\underline{v};\ \underline{r},\ \underline{v})\ d(\Delta\underline{r})d(\Delta\underline{v}) \qquad (55)$$

The integration covers all possible increments, $\Delta\underline{r}$ and
$\Delta\underline{v}$ that may occur. Further, $\psi(\underline{r} - \Delta\underline{r},\ \underline{v} - \Delta\underline{v};\ \Delta\underline{r},\ \Delta\underline{v})$
is the probability that the particle position and
velocity at time t are incremented by $\Delta\underline{r}$ and $\Delta\underline{v}$ amounts
each, within the time interval Δt. This is called the
transition probability. In writing this equation, we have
assumed that the particle velocity at time t + Δt depends
solely on the condition of the particle at time t. Such
an assumption means that the Brownian motion is being
idealized as a one step Markov process. The corresponding
Langevin equation for these particles can also be written
as

$$\frac{dv}{dt} = \beta(\underline{u} - \underline{v}) + \underline{A}\ (t) \qquad (56)$$

209

and over the time interval Δt, we can write

$$\Delta \underline{v} = \beta(\underline{u} - \underline{v})\Delta t + \int_{t}^{t+\Delta t} \underline{A}(\zeta)\, d\zeta \tag{57}$$

$$\Delta \underline{r} = \underline{v}\, \Delta t \tag{58}$$

The stochastic integral $\int_{t}^{t+\Delta t} \underline{A}(\zeta)\, d\zeta$ is the Weiner process and its distribution is deduced by applying Theorem I (Appendix A). Choosing $\psi(t - \zeta) = 1$, the distribution function is seen to be Gaussian and is given as

$$W\{\underline{B}(\Delta t)\} = \frac{1}{[4\pi q \Delta t]^{3/2}} \exp\left[-\frac{\underline{B}.\underline{B}}{4q\Delta t}\right] \tag{59}$$

with

$$\underline{B}\,(\Delta t) = \int_{t}^{t+\Delta t} \underline{A}(\zeta)\, d\zeta \tag{60}$$

The transition probability $\psi(\underline{r}, \underline{v}; \Delta \underline{r}, \Delta \underline{v})$ can be written as

$$\psi(\underline{r}, \underline{v}; \Delta \underline{r}, \Delta \underline{v}) = \psi(\underline{r}, \underline{v}; \Delta \underline{v})\, \delta(\Delta x - v_x \Delta t)\, \delta(\Delta y - v_y \Delta t)\, \delta(\Delta z - v_z \Delta t) \tag{61}$$

indicating that the transition probability can be found from $\psi(\underline{r}, \underline{v}; \Delta \underline{v})$ if Equation (58) holds. Substituting Equation (61) into (55), upon integration over $\Delta \underline{r}$, one has

$$W(\underline{r}, \underline{v}, t+\Delta t) = \int W(\underline{r} - \underline{v}\Delta t, \underline{v} - \Delta \underline{v}, t)\, \psi(\underline{r} - \underline{v}t, \underline{v} - \Delta \underline{v}; \Delta \underline{v})\, d(\Delta \underline{v}) \tag{62}$$

Alternatively, this can also be written as

$$W(\underline{r} + \underline{v}\Delta t, \underline{v}, t+\Delta t) = \int W (\underline{r}, \underline{v} - \Delta \underline{v}, \Delta t)\ \psi(\underline{r}, \underline{v} - \Delta \underline{v}, \Delta \underline{v})\ d\ (\Delta \underline{v}) \qquad (62a)$$

Expanding these probability functions in a Taylor's series, we have [See Chandrasekhar, p. 32-35, for details]

$$\{\frac{\partial}{\partial t} + \underline{v} \cdot \nabla_r W^1 \Delta t + 0\ (\Delta t^2)\} = \Sigma_i \frac{\partial}{\partial v_i}(W \overline{\Delta v_i}) + \frac{1}{2}\Sigma_i \frac{\partial^2}{\partial v_i^2}(W \overline{\Delta v_i^2})$$

$$+ \Sigma_{i<j} \frac{\partial^2}{\partial v_i \partial v_j}(W \overline{\Delta v_i \Delta v_j}) + \ . \ . \ . \qquad (62b)$$

where

$$\overline{\Delta v_i} = \int_{-\infty}^{\infty} \Delta v_i\ \psi\ d\ (\Delta \underline{v}) \qquad (63a)$$

$$\overline{\Delta v_i^2} = \int_{-\infty}^{\infty} \Delta v_i^2\ \psi\ d\ (\Delta \underline{v}) \qquad (63b)$$

$$\overline{\Delta v_i \Delta v_j} = \int_{-\infty}^{\infty} \Delta v_i\ \Delta v_j\ \psi\ d(\Delta \underline{v}) \qquad (63c)$$

The transition probability $(\underline{r}, \underline{v}; \Delta \underline{v})$ is given from Equation (59) as

$$\psi(\underline{v}; \Delta \underline{v}) = (\frac{1}{4\pi q \Delta t})^{3/2}\ \exp\ (- \frac{|\Delta \underline{v} + \beta(\underline{v} - \underline{u})\Delta t|^2}{4 q \Delta t}) \qquad (64)$$

Substituting Equation (64) into (62)-(63), one has

$$\overline{\Delta v_i} = \beta(u_i - v_i) \Delta t \tag{65}$$

$$\overline{\Delta v_i}^2 = 2q\Delta t + 0 \ (\Delta t^2) \tag{66}$$

$$\overline{\Delta v_i \Delta v_j} = 0 \ (\Delta t^2) \tag{67}$$

Equation (62) may be simplified to

$$\frac{\partial W}{\partial t} + \underline{v} \ \nabla_r W = -\beta \underline{u} \ \nabla_r W + \beta \nabla_v \ W \underline{v} + q \nabla_r^2 W \tag{68}$$

which is the Fokker-Planck equation for inertial particles.

When particle inertia may be neglected, particle velocity decouples from the particle position and the Fokker-Planck equation reduces to the standard convective-diffusion equation. When fluid movement is absent, it becomes the ordinary (Fickian) diffusion equation. Recently Lin (1989) has derived the Fokker-Planck equation and the convective diffusion equation for the situation where particle mobility is position dependent (e.g. when a particle is either close to a wall or particle is non-spherical).

The advantage of dealing with the Fokker-Planck equation is that we can consider deposition situations in a rigorous manner. The condition of deposition can be imposed through an absorptive boundary condition at the wall of the collector. Such a treatment on the Langevin equation approach is not possible and one has to rely on a simulational approach.

3. Simulation of the Dynamics of Particle Deposition

Computer simulation of particle deposition, in essence consists of the construction of the trajectories of particles as they move toward a collecting surface. By monitoring these trajectories and from the knowledge of whether or not they will impinge upon the surface and become deposited, both the macroscopic (i.e. total deposition rate or collection efficiency) and microscopic (i.e. local deposition flux and morphology of deposits) information about deposition may be obtained.

As simulation of particle deposition is based on the determination of particle trajectories, the major problems in computer simulation of particle deposition are (a) the efficient solution of the equations of particle motion (b) the identification of the spatial domain through which particle trajectories travel, (c) the specification of the initial conditions of the approachingparticles and (d) the consequences arising from the impaction of particles on the boundaries of the domain. The last three problems are common to all deposition simulations while the specific form of the equations of particle motion to be solved depend upon the fluid flow field as well as the type of operative forces involved. In the following, we shall address the last two problems first followed by a presentation of three specific deposition problems to illustrate the use of the simulation procedure for analyzing various deposition problems.

3.1 Specification of the Spatial Domain, Conditions at Domain Boundaries and Initial Conditions of Particles

By spatial domain, we refer to that part of space in which particle trajectories need to be determined. Generally speaking this portion of space is adjacent to the surface upon which deposition takes place. The boundaries of the domain may be either physical (such as the deposition surface or confining walls) or imaginary as discussed later. The surface through which particles enter the domain is termed the control surface. The position on the control surface occupied by a particle and the velocity the particle possesses when it occupies this position constitute the initial conditions used in determining the trajectory of a particle. The identification of the spatial domain, the consequences when the trajectory of a particle intersects the boundaries and the determination of the initial conditions of the approaching particle at the control surface will be discussed in the following.

3.1.1 Definition of Spatial Domain for Simulation

In general, the types of spatial domains through which flows occur as encountered in deposition problems occur

are either confined, unconfined (i.e. extend to infinity
in some dimensions) or repeating types. When the flow is
confined physically between boundaries, the domain of
interest for simulation coincides with the physical
domain. Flow domains consisting of tubes of circular
cross section can be considered as typical examples.

On the other hand, the flow domains of certain problems
are unconfined. Such domains are obtained as an
idealization of flow situations wherein the physical
boundaries chosen are far away from the deposition
locations such that their effects are insignificant. The
use of a stagnation point flow approximation near the
blunt edge of a body is one such example. In such cases,
the flow domain is a truncated version of the actual flow
where attention is focussed on a small region.

Besides these, one may encounter flows through domains
which consist of a large number of identical repeating
cells. Filter beds may be viewed as assemblages of unit
cells each of which is identical to or is geometrically
similar to the others. An example of such a situation is
the consideration of a fibrous bed as a collection of
fibers which are parallel to each other. In such cases,
the flow domain is chosen to comprise of the unit cell
itself and the flow field within the unit cell is then
determined.

3.1.2 Control Surface & Assignment of Initial Positions

It is obvious that the choice of a control surface is an
important issue. By definition, a control surface is part
of the bounding surface from which all particles under
consideration are enter the flow domain. Consequently,
the extent of the control surface must be chosen to be
large enough to accomodate all those particles which
could be deposited onto the collector. In practice, the
determination of the extent of the control surface
usually requires multiple trials to establish the
boundaries beyond which particles definitely escape
collection.

Initial positions of particles are chosen randomly at the
control surface. If the surface can be described by a
coordinate system, the positions can be generated using a
random number generator. The aim is to obtain a sample of
particle positions which are uniformly distributed
throughout the control surface in accordance with certain
physical requirements. For planar control surfaces with
uniform flow, one may choose the x and y positions from
two independent uniformly random distributions.

3.1.3 Initial Particle velocities

The velocity with which a given particle enters the flow
is part of the required initial condition in the solution

of the quations of particle motion for determining the
trajectory of the particles. For the case when the
Brownian motion effect is negligible, the initial
particle velocity may be assumed to be that of the fluid.
If gravit ational or other forces are operative, the
velocity due to these forces may be added to the host
fluid velocity to obtain the particle velocity.

When particles experience significant Brownian motion, it
is not likely for these particles to possess a definite
velocity upon entry into the flow domain. Instead we must
assume that the initial velocity is a random quantity
with a specified probability distribution. Considering
the particles as being analogous to gas molecules which
are at thermodynamic equilibrium, we may expect their
velocities to be randomly distributed about a mean which
is equal to the velocity of comparable non-diffusive
particles. If the distribution is assumed to be
Maxwellian [see expression given by Equation (70) below],
the initial particle velocity is

$$\underline{v}_o = \underline{u} \text{ at control surface} + \underline{v}_o^r \tag{69}$$

where v is a random quantity which is distributed as
follows:

$$W(\underline{v}^r) = (\frac{m}{2\pi kT})^{3/2} \exp(-\frac{m}{2kT} \underline{v}^r \cdot \underline{v}^r) \tag{70}$$

3.1.4 Boundary Conditions

One of the important problems in developing a computer
simulation of Brownian motion of aerosol particles is to
deal with the movement of particles near either the
physical or imaginary boundaries of the spatial domain.
Physical boundaries represent the solid surfaces impacted
by the particles. Imaginary boundaries are the limits of
the flow beyond which it is assumed that particle
behavior is of no consequence to the deposition problem
at hand. Such boundaries may represent the edges of a
unit cell in the flow domain for example.

Treatment of deposition usually proceeds by assuming that
any particle whose trajectory intersects any solid
surface is assumed to be deposited. The position of
deposition can be approximated by the intersection of the
solid surface with the line connecting the two closest
particle locations on either sides of the collector
surface. Once a particle is deposited, it is normally
assumed to be removed from the flow and thus from future

consideration. It is possible to provide some measure of re-entrainment by modeling the desorption of the particles from various locations on the surface. This aspect has not been investigated in detail.

The assumption of particle removal from the flow upon deposition implies that the deposition surface is a perfectly absorbing wall. For the solution of the corresponding convective diffusion equation, the boundary condition imposed is the vanishing concentration of the diffusing component. (See e. g. Chandrasekhar, 1943), or

$$c_{wall} = 0 \qquad\qquad (71)$$

A more interesting question is the simulation of particle behavior near the imaginary flow boundaries. As remarked above, such boundaries do not correspond to physical boundaries of the flow but are merely idealized representations of the flow domain. In reality, there is a free exchange of particles between the cell and its neighbors across these boundaries. In order to include this exchange in the simulation, we may assume the boundary to be a perfect reflector. A particle is reflected in a specular manner whenever it impacts upon such a boundary in contrast to the deposition condition discussed above. This requirement corresponds to the vanishing of the concentration gradient at the boundary, or

$$D \underline{n}.vc = 0 \qquad\qquad (72)$$

$\underline{n}$ is the unit normal to the collector surface directed outward i.e. into the flow. When the gradient of concentration vanishes, the concentration profile is symmetric about this boundary.

3.1.5 Simulation of deposition at surfaces

The solutions to the Langevin equations described in section (2) can be termed as equilibrium solutions in the sense that they represent the probability distributions for particles that would be diffusing in an unhindered manner. On the other hand, in the proximity of an adsorbing wall (close to the collector surface), particle velocity and displacement distributions are likely to be different from those in the fluid bulk and the difference may be significant. This is because of particles once adsorbed at the surface may not return into the flow. Thus, the distribution of particles (i.e. their concentration) should decrease as they approach the surface. The presence of an adsorbing surface is not accounted for in the solutions of the Langevin equation as discussed hitherto. In obtaining particle deposition fluxes, the probability distributions for particle increments must be suitably altered to reflect this fact.

3.2 Simulation of Brownian particle deposition with negligible particle inertia on single cylindrical collectors (Kanaoka et. al., 1983)

Kanaoka et. al. (1983) studied the deposition of Brownian particles with negligible inertia on a single cylindrical collector as a model depicting fibrous filtration. The Kuwabara cell model was used to represent the flow field and the primary emphasis was placed on a detailed examination of the build-up of particle deposits (i.e. dendritic growth) along the cylindrical collector surface and its effect on deposition rate. (See Figure 1 for details.)

The starting point used by Kanaoka et. al., is the Langevin equation written for the particles in the absence of inertia. The appropriate form of the Langevin equation and its solution is discussed in section 2. (See case a.i.) The path of Brownian particles introduced into the flow was simulated by Kanaoka et al. by utilizing these solutions. They assumed that the fluid motion can be well represented by a constant over a short time interval. Then, the particle velocity can be replaced by the relative velocity between the particle and the fluid. The position of the particle after time t can then be given by the solution presented in section 2.b.i [Equations (17) and (18)].

This equation may be rewritten in a discrete form as

$$\underline{r}_i(t+\Delta t) = \underline{r}_{i-1}(t) + \underline{u}(t)\ t\Delta + 3\underline{r}^r \tag{73}$$

where $\underline{r}_i(t+\Delta t)$ is the particle's position vector at time , $t+\Delta t$ $\underline{r}_{i-1}$ is the particle's position vector at time t , $\underline{u}\Delta t$ is the incremental displacement in the particle's position due to the convective motion of the fluid and $\underline{r}^r$ is the random vector representing the change in particle's position owing to its Brownian motion. Based on the assumption that the flow field is given by the Kuwabara model (Kuwabara, 1959), the fluid velocity $\underline{u}$ is expressed as

$$\underline{u} = (\frac{\partial \psi_s}{\partial y}, -\frac{\partial \psi_s}{\partial x}, 0) \tag{74}$$

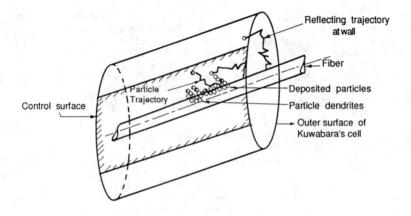

Figure 1   Schematic diagram of collection by fibers. Simulation of Kanaoka et.al.

and the stream function, ψ_s is given as

$$\psi_s = \frac{y}{2p} [(1 - \frac{\alpha_p}{2}) \frac{1}{x^2+y^2} - (1-\alpha) + \ln (x^2+y^2) - \frac{\alpha_p}{2} (x^2+y^2)] \tag{75}$$

$$\alpha_p = -\ln \rho + \rho - \frac{\rho^2}{4} - 3/4 \tag{76}$$

where x and y are the Cartesian coordinates and α_p is the packing density of the filter. The spatial domain for particle simulation consists of a hydrodynamic cell enclosing a unit length of the fiber. The distribution of r_r is Gaussian and is given by Equation (18).

These equations can be used to simulate the trajectory of a Brownian particle in a given flow field if r^{i-1} and u^{i-1} are known. Initially a Brownian particle is introduced at the control surface, with its position and velocity specified by the methods described in section 3.1.2. The position of the Brownian particle at the end of a time increment is predicted by using Equation (73). By repeating this process for a new time step with u readjusted to the fluid velocity at the particle's current position, we can construct a chain of displacements which will represent a sample trajectory of the particle. This procedure is carried out until the particle either moves out of the flow domain or impinges on the collector's surface. A flow chart demonstrating this simulation scheme is shown in Figure 2. Using this method described above, Kanaoka et. al. (1983) examine the particle deposition process and also the effect of particle deposition on subsequent collection by the fibers.

An important question arises when the Kuwabara flow field is used throughout the entie deposition process when the extent of deposition becomes significant. In reality, one would expect the fluid velocity near the collector to be affected by the presence of deposited particles. Thus, an approaching particle moving toward a cylindrical collector already with substantial particle deposits will experience a different flow field than that given by Equations (75) & (76). The use of the assumption that there is on changes in fluid flow field is necessitated by a lack of practical means to predict the changes to the flow field caused by the presence of particle deposits.

In carrying out their simulation, Kanaoka et. al. first

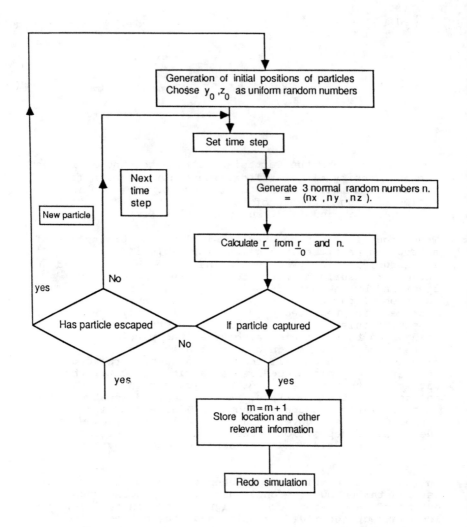

Figure 2   Flow chart of simulation of Kanaoka et.al.

generated the initial positions of particles at the control surface and particle trajectories were then determined from Equation (73) and their outcomes were monitored. An inventory of the number of deposited particles and their deposition positions was maintained continuously from which the number of particle dendrites formed and the number density of dendrites consisting of specific numbers of particles were collected.

During the initial stages of deposition, collection of particles by the cylindrical collector itself (i.e. primary collection) was dominant. As time progressed, collection of individual particles by previously deposited particles became more important. This stage was marked by the development of dendritic agglomerates which grew outward from the fiber. Even at relatively high values of the Peclet number (Pe = 200), deposition was found to persist at the rear of the fiber which is in agreement with the hypothesis proposed earlier by Payatakes & Tien (1976, 1980). Kanaoka et al.'s results also indicated that maximum deposition of particles occurs at an angle of approximately 20 away from the front stagnation point of the cylindrical collectors. Further, this maximum shifted to larger angles as the Peclet number increased. Since in the absence of flow, diffusional deposition is uniform on the fiber surfaces, this maximum is a consequence of convective transport alone.

An interesting result of this work was that the number density of dendrites consisting of larger number of particles such as 6, 7, or 8 was significantly lower than those dendrites composed of relatively fewer particles (i.e. up to 3). This trend was found to persist with time. The growth rate of an average dendrite size was found to depend primarily on the relative size of the fibers to the particles (N_R). It was largely independent of the Peclet number.

Information on the instantaneous collection efficiency was also obtained from the simulation. The collection efficiency for the cylindrical collector (the same as the single fiber efficiency of the filter) is given from the average number of particles that escape capture from the cylindrical collectors in the time interval between the deposition of two particles. Let $<M>$ be the average number of particles that have escaped after the mth particle has deposited and before the m+1 th particle has deposited.

$$\eta = 1/<M> \tag{77}$$

with the average being taken over a number of simulation runs corresponding to the same initial conditions. The instantaneous collection efficiency can be related to the extent of particle deposition expressed in terms of mass of deposits m as

$$\eta/\eta_o = 1 + am^b \qquad (78)$$

where η_o is the efficiency of the cylindrical collector at the beginning of deposition and m, the mass of particles deposited per unit volume of the bed. By definition, m is given as

$$m = (4/3) \; \rho_p \alpha_p M \; N_R^3 \qquad (79)$$

where ρ_p is the particle density, M is the number of deposited particles per unit length and α_p the fiber packing density of the filter. N_R is defined as a_p/a_f where a_p and a_f are particle and fiber radii. From the simulation results, values of η at various values m were determined first. Then they were correlated in the form of Equation (78). The values of a and b obtained by Kanaoka et al. are listed in Table 1. Since the values of b were found close to unity, the correlation can be modified as

$$\eta/\eta_o = 1 + \lambda m \qquad (80)$$

The values of λ are also presented in Table 1 for two different particle sizes. λ was found to vary from 15 to 0.8 with $N_R = 0.1$ and from 3.1 to 0.6 with $N_R=0.2$ suggesting that the collection efficiency increases faster with particle loadings at high Peclet number. This conclusion was consistent with the results obtained if inertial impaction is the dominant mechanism of deposition as observed by Kanaoka et al. (1980).

3.3 Brownian Particle Deposition on Single Spherical Collectors-Dynamics in Granular Filters: Simulation of Gupta and Peters (1984, 1985, 1986).

Gupta and Peters (1984, 1985, 1986) developed a method for simulating the deposition of particles on single spherical collectors with both the particle inertia and Brownian effects being significant and assuming that the fluid velocity around the single collector is given by

TABLE 1

Values of constants a and b in Equation (78) and λ in Equation (80) as determined by Kanaoka et. al. (1983). These constants determine the increase in single fiber collection eficiency with particle loading in a fibrous filter.

Pe	a		b		λ	
	N_R = 0.1	0.2	0.1	0.2	0.1	0.2
200	0.614	0.599	1.29	0.991	0.789	0.588
1000	2.15	1.02	1.03	1.03	2.19	1.07
5000	3.84	1.13	1.00	1.14	3.84	1.42
α	–	–	–	–	14.8[*]	3.09[*]

[*] Obtained from Kanaoka et. al. (1980).

TABLE 2

Details of Simulation, Showing Number of Particles Deposited at Each Time Step. Air at 20°C, d_p = 0.05 μm, u_∞ = 10 cm/s, H = 1 cm, F_p = 0.1, Total number of particles considered 1000.

Time	Δt	N	n_{dep}	Total deposited
0.5	0.1	1000	0	0
0.6	0.1	1000	63	63
0.7[*]	0.1	937	244	--
0.65[*]	0.05	937	141	--
0.625	0.025	937	67	130
0.65	0.025	870	81	211
0.675[*]	0.025	789	104	--
0.6625	0.0125	789	64	275
0.675	0.0125	725	50	325
0.6875	0.0125	675	61	386
0.7000	0.0125	614	50	436

223

Stokes' solution. In the following, we will detail their simulation procedure, its application as shown by Gupta & Peters, and their results on deposition fluxes.

Consider a Brownian particle in a fluid stream over a short time Δt over which we may assume that the fluid velocity and the external force experienced by the particle may be constant. The dynamics of such a particle are given by the Langevin equation as developed in section 2.1.2.2. Notice that the vectors R_1 and R_2 are stochastic and their joint distribution is given by equations (23-26). The fluid velocity according to Stokes' solution is

$$u_x = u_\infty \left[\frac{3}{4} \frac{Rzx}{r^3} \left(\frac{R^2}{r^2} - 1 \right) \right] \tag{81}$$

$$u_y = u_\infty \left[\frac{3}{4} \frac{Rzy}{r^3} \left(\frac{R^2}{r^2} - 1 \right) \right] \tag{82}$$

$$u_z = u_\infty \left[\frac{3}{4} \frac{Rz^2}{r^3} \left(\frac{R^2}{r^2} - 1 \right) + 1 - \frac{1}{4} \frac{R}{v} \left(3 + \frac{R^2}{r^2} \right) \right] \tag{83}$$

where u_∞ is the free stream velocity of the fluid oriented in the z direction, r is the radial position, R is the radius of the collector sphere which was placed at the origin of the coordinate system.

The simulation of particle trajectories can be conducted in the same manner as in the previous case. Figure 3 for a depicts a schematic description of the fluid flow and the control surface where particles are assumed to enter the flow. The algorithm used for the simulation is illustrated in Figure 4. The random deviates R_1 and R_2 can be sampled by employing standard random number generating software [See Appendix B].

The deposition flux predicted by the simulation can be determined from the limiting trajectory, namely, that trajectory beyond which particles do not deposit. Due to the Brownian motion of the particles, this limiting trajectory is defined after averaging over a number of simulation trials under corresponding initial conditions. Gupta & Peters obtained the deposition flux over a range of particle Stokes numbering varying from very small

224

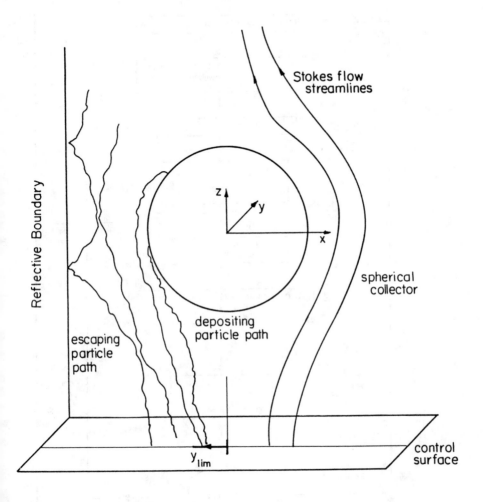

Figure 3   Schematic diagram for deposition on spherical collectors.
(Gupta & Peters method)

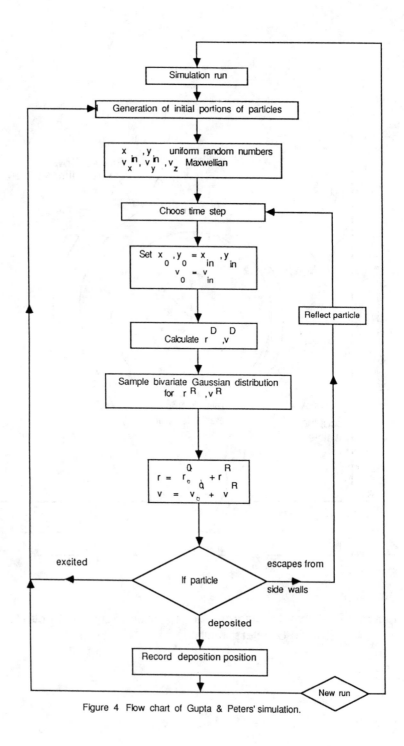

Figure 4 Flow chart of Gupta & Peters' simulation.

values to large values (> 1.0). When the particle
Stokes number are small, the inertial effects experienced
by the particles are insignificant compared to their
diffusive motion. The particles may indeed be considered
to behave like molecules of a gas species diffusing to
the collector's surface. The simulation results of Gupta
& Peters matched the predcition of deposition fluxes
using the convective diffusion equation for a diffusing
molecular species obtained by Levich (1962). Whenthe
particle Stokes' numbers are large, diffusional effects
are negligible and deposition is governed primarily by
deterministic mechanisms such as inertial impaction,
interception and gravitational attraction. The deposition
flux results of Gupta & Peters matched those of Langmuir
accurately under this regime.

When both diffusion and particle inertia are dominant
mechanism affecting deposition (i.e. in the range of
intermediate Stokes' numbers), the results of Gupta &
Peters agreed very well with the theoretical predictions
of Fernandez de la Mora & Rosner (1981. 1982) based on
the two fluid theory.

It is evident that the Brownian dynamics simulation
method based on the solutions of the Langevin equation is
a powerful tool in solving deposition problems where a
combination of inertial and diffusive mechanisms needs to
be considered. Specifically, the results of Gupta and
Peters are useful in coating problems and in analyzing
granular filtration of aerosols. An important application
of the Brownian dynamics method is in studying the
effects of hydrodynamic interactions between aerosol
particles. This application arises in considering the
dynamics of a relatively concentrated aerosol where the
motion of each of the particles can no longer be
considered as independent but is coupled with its
neighbors. In such cases, hydrodynamic interaction force
between a particle and its neighbors must be included in
examining the particle motion. With much of the recent
advances in the understanding of the hydrodynamic
interactions of particles, incorporation of such effects
into the Brownian dynamics simulation is made possible.

The Brownian dynamics method relies on simulating the
trajectories of particles within the flow. Since the
trajectories are described in a stochastic sense, one has
to obtain large enough numbers of the samples of the
trajectories to obtain meaningful deposition results.
Further, the trajectories themselves contain sequences of
large numbers of steps. It can be foreseen that
computations are likely to be cumbersome and numerous
even in relatively simple flow situations. An application
of a variant of this method which shows a way of avoiding
cumbersome calculations in linear situations is shown
next.

3.4 Browian Particle Deposition in Stagnation Point Flow

As a final example of applying the simulation technique , the problem of the deposition of Brownian particles in stagnation point flow will be considered. Stagnation point flow is often realized in practical situations such as flow over the nose of a blunt body or that over a workbench in clean rooms.

Referring to Figure 5, the spatial domain of interest is described as a large rectangular region enclosing the flow. For purposes of analysis, the momentum boundary layer present in the immediate neighborhood of the deposition plane is ignored. A gaseous stream enters the flow domain at the top along a horizontal plane situated at y = H, with a velocity U_∞. Particles may enter this flow at any location x = x_i all along this plane. The flow streamlines are rectangular hyperbolas with the fluid exiting along both the side walls.

The flow field in stagnation point flow is given as

$$\alpha x \tag{27a}$$

$$\underline{u} \quad = $$

$$-\alpha y \tag{27b}$$

The equation of motion of a Brownian particle can be described using Equations (28) and (30). The solution is given by Equations (30) thru (34).

The particle paths can be simulated using the methods presented earlier. The control surface is the line y=H along which particles may be assumed to enter at any location x_i. The initial velocities of the particles may be determined as shown in section 3.1. The position & velocity of a single particle after elapse of time t can be calculated by applying equations (30) & (34). The random deviates $\underline{R}_1$ & $\underline{R}_2$ are generated using random number generators described in Appendix B.

The determination of the deposition flux in the stagnation point flow case is simplified by the fact that an exact solution to the Langevin equation is available. However, a simulational approach is still necessary because of the following complication. If the deposition plane was absent, one could choose an arbitrarily large time step so that the particles which started at the initial position arrive near the x axis, (y=0) within the single time step itself. When the deposition plane is present though, it serves to adsorb particles and so remove them from the gas phase. The use of Equations

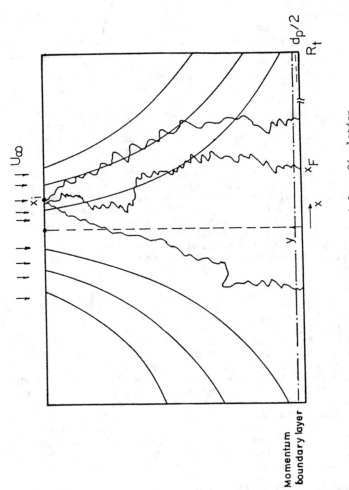

Figure 5   Particle Path Types Obtained from Simulation

(23) - (38) to represent the probability density of particles will not be valid. Thus, if the particles are far away from the deposition surface, the probability distributions of the particle positions after taking n time steps each of a duration t and the distributions obtained by a single interval nt would be identical. However, when significant numbers of particles under consideration begin to deposit, one must modify the distributions. To account for this, one begins simulation with a set of particles at a certain location with a given initial velocity at y=H. As time progresses, these particle will form a particle cloud with increasing size, and disperse through the spatial domain because of random Brownian motion. When this set of particles comes near the deposition plane and starts to deposit onto the plane, those deposited particles are then removed from the initial set. This principle is illustrated with a schematic of the simulation procedure for studying deposition in stagnation point flows in Figure 6.

The deposition flux J, may be determined from conducting the simulation as follows. A large number of particles whos initial positions are randomly distributed all along the control surface at y = H are chosen. The trajectories of these particles are tracked over time intervals which progressively become shorter as the deposition plane is approached. The deposition flux may be determined from a consideration of the number of particles which have deposited at given location. This ad-hoc procedure to determine deposition fluxes gives results [as shown by Ramarao & Tien (1990b)] which agree well with a more accurate procedure consisting of integrating the probability of deposition of a particle over the control surface.

The deposition situation is general, we may obtain a local deposition flux which is dependent upon the location on the deposition plane and also an average deposition flux which is an average of the local deposition flux over the whole of the deposition plane. For the case of convective diffusion in stagnation flow, these two fluxes are equal as shown by Yuu & Jotaki (1979). In Table 3, the average deposition flux over various intervals along the deposition plane obtained from simulation of two particle sizes are shown. These results indicate that J is essentially independent of s_f. The results for the 0.3 μm particle case seem to suggest that this lack of dependence on x_f can also be expected for particles of large size when both the deterministic and stochastic force are equally important.

Particle deposition fluxes were calculated by Ramarao & Tien (1990b) for a variety of sample situations are presented in Figure 7. In this figure, value of J corresponding to larger particles and three different flow fields (H=1, 10 and 100 cm) vs d are shown. The simulation results are mean value samples including up to 20 trials. In addition, values of J according to

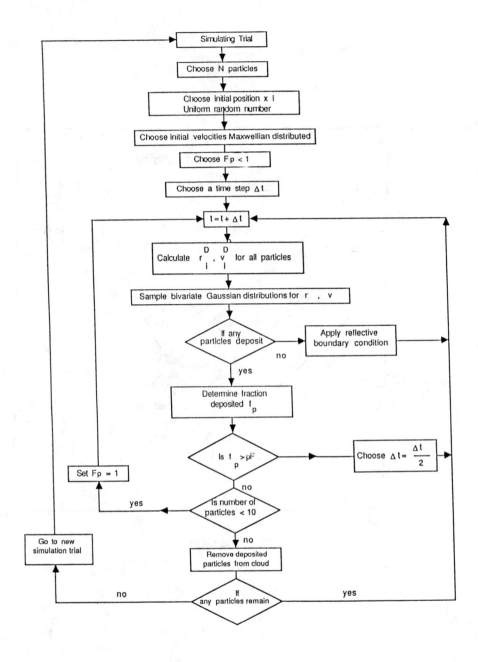

Figure 6 Simulation flow chart for ideal stagnation paint flow

231

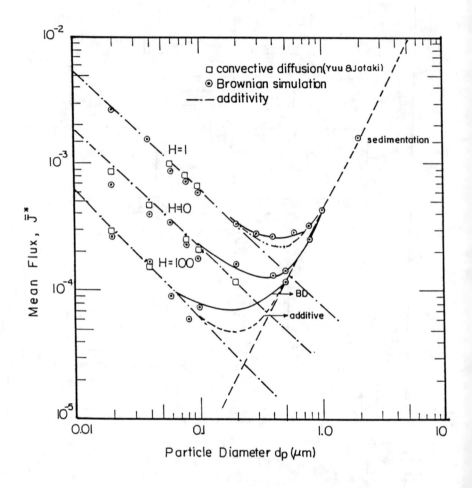

Figure 7. Deposition flux for stagnation flow.

Brownian diffusion (i.e. the solution of Yuu and Jotaki) and sedimentation alone as well as those combinations are also shown. The sedimentation results were obtained using particle terminal velocities calculated from Stokes law. The simulation results agree with either the convective diffusion or sedimentation results for small or large diameters. For d_p between 0.1 and 0.5 um, these differences become substantial. It is also clear that predictions based on the addition approximation tend to under predict J for particles in this size range.

It is also of interest to compare the details of the method of calculation based on the simulation results with the Brownian dynamics simulation method devised by Gupta and Peters. The simulation method outlined in section 3.2 can be used to determine particle fluxes in stagnation point flow also. In this case, we replace the flow field of Equations (81) - (83) by the ideal stagnation flow field of Equation (27). We shall refer to the simulation method outlined in this section (3.3) the linear solution method (LS) denoting that we are solving the Linear Langevin equation. The linear solution gives us the probability density of particle position and velocity components after an arbitrary time elapsed, t.

A comparison between the results of the two simulation methods may be made by looking at the means and variances of particle positions after a time interval t has elapsed. Supposing we start with a set of particles whose initial positions were chosen at random along the control surface (x_i of Figure 5), with an initial velocity v after the elapse of a certain time t, we will observe an expanding particle cloud in the air with the particles moving into the flow field. The position vector distribution of the particle displacements will be Gaussian about certain mean values. In principle, this cloud of particles must show the same distribution in position vectors and the velocities irrespective of whether we used the Brownian dynamics simulation method of Gupta and Peters or the linear solution obtained in this work. The comparison of the mean and variance in the x and y components of the particle corresponding to a given solution obtained by using these two methods is presented in Table 4. Notice that both the means and variances match very well suggesting the close correspondence between the simulation method of Gupta and Peters and our solution.

A comparison of the total CPU time of a VAX 8550 computer for determining particle flux corresponding to specified conditions by considering a total of 1000 particles and making 10 trials according to the method of Gupta and Peters is presented in Table 5. It is clear that the time requirement increases rapidly as the initial height, H increases. On the other hand, the time requirement based on the linear solution for 2000 particles was 242s and independent of H. The practical advantage afforded by the linear solution is obvious.

233

TABLE 3

Variation of Deposition Flux J^* Calculated from Equation
(73) with Location of Interval $[^XF_1, {}^XF_2]$ on Deposition
Plane. H = 1 cm, u = 10 cm/s, air at 0°C.

$[^XF_1, {}^XF_2]$	$J^*(\Delta x_F)$	
	$d_p = 0.01 \ \mu m$	$d_p = 0.3 \ \mu m$
0.0-1.0	0.0051	0.00026
1.0-2.0	0.0048	0.00028
2.0-3.0	0.0049	0.00024
3.0-4.0	0.0050	0.00023
9.0-10.0	0.0046	0.00029

TABLE 4

Position Statistics of a Particle Cloud which Starts at
an Initial Position, x_i. Comparison of Means and
Variances in x and y Coordinates of Particles as
Predicted by Simulations of Gupta and Peters (GP) and
Using Linear Solution (LS). Air at 0°C. Total Number of
Particles is 500, Time Interval Use4d in Gupta and Peters
Simulation is 0.01s, x_i = 0.1 cm, d_p = 0.01 μm, H = 100
cm, u∞ = 10 cm/s.

Particle Position

Time,s	Method	Mean, cm		Variance, cm^2	
		$\bar{x}$	$\bar{y}$	σ_x^2 $\times 10^{-3}$	σ_y^2 $\times 10^{-3}$
0.1	GP	0.10037	99.0051	0.1114	0.1067
	LS	0.10098	99.0049	0.1040	0.1060
1.0	GP	0.11023	90.478	0.2154	0.9718
	LS	0.11042	90.483	0.1140	0.9699
2.0	GP	0.12410	81.866	0.2563	0.1710
	LS	0.12200	81.873	0.2534	0.1764
3.0	GP	0.1369	74.072	0.4161	0.2197
	LS	0.1348	74.081	0.4235	0.2414
4.0	GP	0.1510	67.017	0.5964	0.2488
	LS	0.1489	67.032	0.6313	0.2947

Fernandez de la Mora & Rosner predicted an interesting inertial enrichment effect of the particle concentrations as an aerosol flow approaches an obstacle. As is well known, when particle inertia is significant, the particle trajectories deviate from the streamlines of the fluid by significant extents. By assuming that the particles in the aerosol formed a continuum and by analyzing the equation of continuity for the gas and particulate phases, Fernandez de la Mora & Rosner predicted that the particle concentrations would increase if inertia was a dominant force acting on the particles. This enrichment of particle concentrations increases the flux of particles due to diffusion at locations adjacent to the deposition surface. Thus it is an important effect. The Brownian dynamics simulation method we have described above can be exploited well to yield particle concentrations at any location within the flow also. Results on the inertial enrichment of particle concentrations in an aerosol in ideal stagnation point flow were recently reported by Peters et. al. (1989). In this investigation, Peters et. al. used the two fluid approach of Fernandez de la Mora & Rosner effectively. Ramarao & Tien (1990b) also studied the concentrations of particles in a two dimensional ideal stagnation point flow using the principles of the Brownian dynamics simulation method. Increases in particle concentration close to the deposition surface were found by them. The predicted inertial enrichment effect was found to agree very well with the predictions obtained using the two fluid theory of Fernandez de la Mora & Rosner (Peters et. al. 1989).

TABLE 5

Simulation Time vs. H, Method of Gupta and Peters
d_p = 0.01 μm, Δt = 0.01 s, Total Number of Particles Considered, 1,000 Number of Trials, 10. Time requirement for the linear solution was 242 independent of H.

H(cm)	CPU Time (s)
1	210
2	435
5	1106
8	1829
10	2333
20	5142
50	12847
100	27105

4. Summary

The dynamics of diffusive aerosols can be effectively studied from the standpoint of the Langevin equation. The Langevin equation furnishes us the added flexibility of being able to deal with particle inertia in conjunction with the Brownian diffusive motion of the aerosol particles using fundamental descriptions from theoretical physics. Furthermore, the Langevin equation provides us the ability to deal with non-steady forms of the hydrodynamic drag on the particles. These advantages are not available in the other methods of determining deposition fluxes, namely, using the two fluid theories or using ad hoc approximations to the convective diffusion equation to account for particle inertia.

In many instances of aerosol flows, one is interested in the deposition of the particles from the gas phase onto the surfaces of solid objects. Such deposition problems can be studied using the Brownian dynamics simulation methods in various forms as shown in this chapter. It is also possible to derive a generalized Fokker-Planck equation from each form of the Langevin equation that has been discussed in this chapter. When particle inertia can be neglected, the Fokker-Planck equation reduces to the convective diffusion equation. Under this condition, it is possible to determine particle deposition fluxes onto solid surfaces by solving the convective diffusion equation. Solution of the Fokker-Planck equation when inertia is present is not a simple task. Although it is possible to obtain probability densities in specialized cases when the Fokker-Planck equation is linear, it is difficult to use them to predict particle deposition fluxes. Thus, the Brownian dynamics method of simulating particle trajectories and their deposition becomes invaluable in tackling such problems.

Nomenclature

A	Stochastic acceleration term
b	Exponent, constant in various equations
c	Concentration of particles
F	Forces on particles
F, G, H	Components of covariance matrix
J	Particle deposition flux
k	Bolzmann constant
m	Particle mas, mass of particles loaded on filter
n	Unit normal vector
p	packing density of fibrous filter
q	$\beta\, kT/m$
r	Position vector
R	Random variable
T	Absolute temperature of gas, aerosol
t	Time
u	Fluid velocity
v	Particle velocity
W	Probability density function
x, y, z	Coordinate axes

Subscripts

d	Drag
e	External
o	Initial value
i	A coordinate axis
p	Particle

Superscripts

r	Random quantity

Greek Letters

α	Flow field constant in stagnation flow
α_p	Kuwabara factor
β	Friction coefficient
δ	Dirac delta function; also used to define equation for x and y axes
μ	Shear viscosity of fluid, also used to denote various constants in equations
υ	Kinematic viscosity of fluid
ψ	Transitional probability
ψ_s	Stream Function
η	Collection efficiency
ρ_p	Particle density

References

Arminski, L. and S. Weinbaum. 'Effect of waveform and duration of impulse on the solution to the Basset-Langevin equation,' Phys. Fluids, 22, 3, 404-411 (1979).

Chandrasekhar, S. 'Stochastic problems in physics and astronomy,' Rev. Mod. Phys., 15, 1, 1-91 (1943).

Corrsin, S. and J. Lumley. 'On the equation of motion for a particle in turbulent fluid,' Appl. Sci. Res. A6, 114 (1956).

Fernandez de la Mora, J. and D. E. Rosner. 'Inertial deposition of particles revisited and extended: Eulerian approach to a traditionally Lagrangian problem,' PCH PhysicoChemical Hydrodynamics, 2, 1, 1-21 (1981).

Fernandez de la Mora, J. and D. E. Rosner. 'Effects of inertia on the diffusional deposition of small particles to spheres and cylinders at low Reynolds numbers,' J. Fluid Mech., 125, 379-395 (1982).

Friedlander, S. K. 'Smoke, Dust and Haze,' Wiley & Sons, New York, NY (1977).

Gupta, D. and M. H. Peters. 'A Brownian dynamics simulation of aerosol deposition onto spherical collectors,' J. Coll. Interf. Sci., 104, 2, 375-389 (1985).

Gupta, D. and M. H. Peters. 'On the angular dependence of aerosol diffusional deposition onto spheres,' J. Coll. Interf. Sci., 110, 1, 286-291 (1986); 110, 1, 301-303 (1986).

Hinch, E. J., 'Application of the Langevin equation to fluid suspension,' J. Fluid Mech., 72, 3, 499-511 (1975).

Hinze, J. O. 'Turbulence,' McGraw-Hill, New York NY (1975).

IMSL International Mathematical and Statistical Libraries, Version 9, Houston, TX.

Kanaoka, C., H. Emi and T. Myojo. 'Simulation of the growing process of a particle dendrite and evaluation of a single fiber collection efficiency with dust load,' J. Aerosol Sci., 11, 377 (1980).

Kanaoka, C., H. Emi and W. Tanthapanichakoon. 'Convective diffusional deposition and collection efficiency of aerosol on a dust loaded filter,' AIChE J., 29, 6, 895-902 (1983).

Langmuir, I. J. Meteorol., 5, 175 (1948).

Levich, V. G. 'Physico-Chemical Hydrodynamics,' Prentice Hall, Englewood Cliffs NJ (1962).

Lin, S. P. 'Convective diffusion in position-dependent drag fields,' J. Coll. Interf. Sci., 131, 1, 211-217 (1989).

Papoulis, A. 'Probability, Random variables and Stochastic processes,' McGraw-Hill, New York (1965).

Payatakes, A. C. and C. Tien. 'Dendritic deposition of aerosols by convective Brownian diffusion for small, intermediate and high particle Knudsen numbers,' AIChE J., 26, 443 (1980).

Peters, M. H. and D. Gupta. 'Brownian dynamics in fixed and fluidized bed filtration,' AIChE Symposium Series, 234, 80, 98-105 (1984).

Peters, M. H. and D. W. Cooper. 'Approximate analytical solutions for particle deposition in viscous stagnation-point flow in the inertial-diffusion regime with external forces,' IBM Research Report RC 14509 (#64971), IBM Research Division, Yorktown Heights, NY (1989).

Maisel, H. and G. Gnugnoli. 'Simulation of discrete stochastic systems' Science Research Associates, Inc., Chicago, IL (1972).

Ramarao, B. V. and C. Tien. 'Role of Basset force in particle deposition in stagnation flows,' to appear in J. Aerosol Sci. (1990a).

Ramarao, B. V. and C. Tien. 'Aerosol particle deposition in two dimensional stagnation point flows,' J. Aerosol Sci., 20, 7, 775-785 (1989).

Ramarao, B. V. and C. Tien. 'Diffusional deposition of aerosols in stagnation point flows.' (1990b).

Ramarao, B. V. and C. Tien. 'Diffusional behavior of small particles: On the solution of the Basset-Langevin equation,' (1990c).

Tchen, C. M. Doctoral Dissertation, Technical University, Delft, Netherlands (1947).

Yuu, S. and T. Jotaki. 'The calculation of particle deposition efficiency due to inertia, diffusion and interception in a plane stagnation flow,' Chem. Engng. Sci., 33, 971-978 (1978).

Appendix A

In this appendix, we will outline two theorems which are of use to determine the probability distributions of integrals involving stochastic kernels. Specifically, the stochastic kernel we shall be concerned with is a Brownian motion i.e. a white noise Gaussian process. Theorem I:

Let

$$R = \int_0^t \psi(t-\xi)A(\xi)\,d\xi \tag{A1}$$

where $A(\cdot)$ is the stochastic white noise process with the following properties:

$$<A(t)> = 0 \tag{A2}$$

$$<A(t) \cdot A(t+\tau)> = K\delta(t-\tau) \tag{A3}$$

Then the probability distribution of $\underline{R}$ is given by

$$W(R) = \frac{1}{\left[4\pi q \int_0^t \psi^2(t-\xi)\,d\xi \right]^{3/2}} \quad \exp\left[-\frac{R \cdot R}{\int_0^t 4q\psi^2(t-\xi)\,d\xi} \right] \tag{A4}$$

The proof follows from Lemma I given in Chandrasekhar (1943) p.23-24. The slight deviation here is the use of the convolution integral with the function $\psi(t-\xi)$.

Theorem II.

Let

$$R = \int_0^t \psi(t-\xi)A(\xi)\,d\xi \tag{A5}$$

240

$$S = \int_0^t \phi(t-\xi) A(\xi) \, d\xi \qquad \text{(A6)}$$

Then the joint probability distribution of $\underline{R}$ and $\underline{S}$ is

$$W(R,S) = \frac{1}{8\pi^3 (FG-H^2)^{3/2}} \exp\left[-\frac{GR \cdot R - 2HR \cdot S + FS \cdot S}{2(FG-H^2)} \right] \qquad \text{(A7)}$$

where

$$F = 2q \int_0^t \psi^2(t-\xi) \, d\xi \qquad \text{(A8)}$$

$$G = 2q \int_0^t \phi^2(t-\xi) \, d\xi \qquad \text{(A9)}$$

$$H = 2q \int_0^t \phi(t-\xi) \psi(t-\xi) \, d\xi \qquad \text{(A10)}$$

The proof of this theorem is also given in Chandrasekhar (1943, p. 26-27). Note that R and S are bivariate Gaussian distributed. The important restriction placed on ψ and ϕ in this as well as the former theorems are that they be slowly varying functions of time as compared with A(t).

241

Appendix B

The generation of random numbers by using the digital computer is described in this appendix. Generation of a uniformly distributed random deviate can be accomplished by a large number of random number generating algorithms. Computer generated random numbers are referred to as pseudorandom numbers because a particular starting (seed) number always gives the same random number according to the algorithms employed. Moreover, all computer generated random sequences possess a finite length and they will tend to cycle after a large interval. One must keep sight of these two facts when using random number generators in computer simulations.

1. Uniform Random Number Generator

One of the most basic algorithms is a uniform random number generator. This algorithm produces a random deviate uniformly distributed within the closed interval [0,1] such that the probability density $p(x)$ of obtaining x within an infinitesimal interval centered about x is a constant value independent of x. We provide a sample linear congruential method for generating uniformly distributed random numbers in Figure 8. This method relies on the generation of two sequences of random deviates according to the expressions

$$X_{n+1} = (cX_n + a) \bmod m \qquad\qquad\qquad (B1)$$

$$Y_{n+1} = (gY_n + h) \bmod m \qquad\qquad\qquad (B2)$$

where X_n and Y_n are two seed numbers. Other random number generators can also be found in standard statistical and mathematical software packages such as IMSL. IMSL utilizes a congruent method for generating random numbers. Let S be a seed number. The set of uniformly distributed random numbers R (i=1,...n) can be generated by using the following equations.

$$S_i = 7^5 S_{i-1} [\bmod 2^{31} - 1] \qquad\qquad (B3)$$

$$R_i = 2^{-31} S_i \qquad\qquad\qquad\qquad (B4)$$

R_i is the output uniform random numbers, S_o is the seed, i is an iteration number.

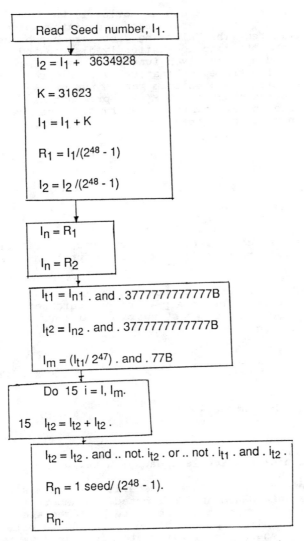

Figure 8 Flow chart for a single uniform random number generator.
(Maisel & Gnugnoli, 1972, p. 432-433).

2. Generation of non-uniform random deviates

In many instances where Brownian motion is to be simulated, we have to generate random deviates which are normal distributed. The following procedure can be used in general for obtaining non-uniformly distributed deviates according to a given functional form. The procedure is illustrated schematically in Figure 9. We first generate standard uniform random numbers R $(i=1,\ldots n)$ as shown above. If $F(x)$ represents the cumulative probability distribution function of the distribution that is desired, then u_1 is a uniform random number,

$$F(x) = \int_0^x b(x)\,dx \tag{B5}$$

$$x = F^{-1}(u) \tag{B6}$$

Thus the inverse of $F(x)$ obeys the desired distribution. $F(x)$ must be a monotonically increasing function of x for this procedure.

A simple method suggested by Box and Muller (1958) involves the generation of a pair of standardized normal numbers, x_1 and x_2 from two standard uniform numbers, u_1 and u_2 by letting

$$x_1 = \sqrt{-2 \ln u_1}\,\cos(2\pi u_2) \tag{B7}$$

$$x_2 = \sqrt{-2 \ln u_1}\,\sin(2\pi u_2) \tag{B8}$$

3. Generation of Multivariate Gaussian Deviates

Finally, let us consider the generation of vectors which are multivariate Gaussian distributed. A number of statistical and mathematical software packages are available for this purpose. In the following we will describe the generation of bivariate Gaussian deviates using the procedure followed by the IMSL subroutine GGNSM. Suppose that S is the covariance matrix of the bivariate Gaussian distribution and also let us assume that the mean of this distribution is zero. We can factor S into its LU components as

$$S = L \cdot L^T \tag{B9}$$

$$L = \begin{bmatrix} a_{11} & 0 \\ a_{21} & a_{22} \end{bmatrix} \tag{B10}$$

244

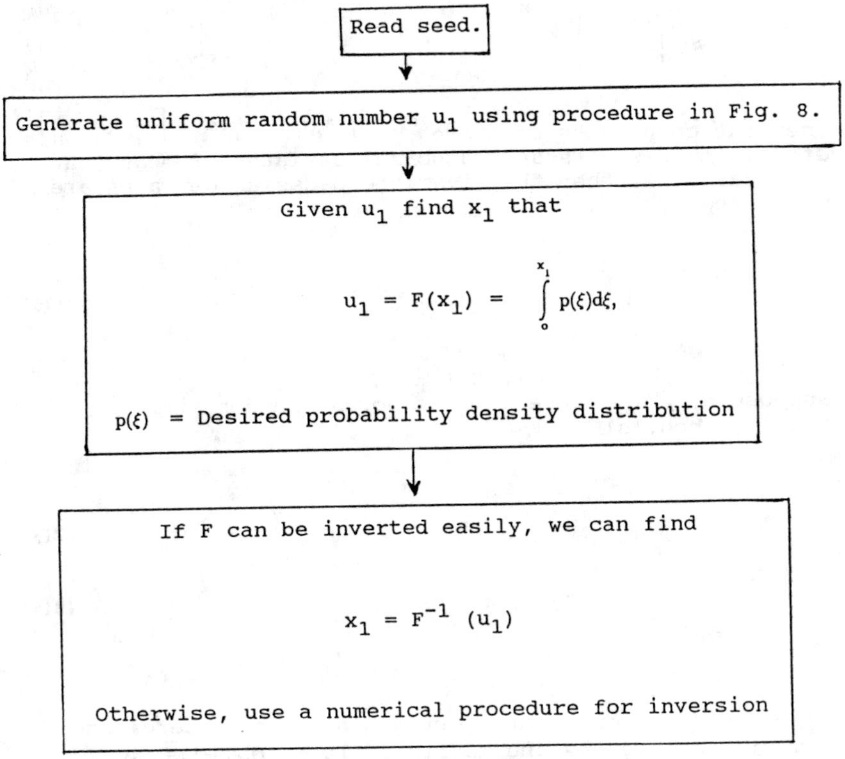

Read seed.

Generate uniform random number u_1 using procedure in Fig. 8.

Given u_1 find x_1 that

$$u_1 = F(x_1) = \int_o^{x_1} p(\xi)d\xi,$$

$p(\xi)$ = Desired probability density distribution

If F can be inverted easily, we can find

$$x_1 = F^{-1}(u_1)$$

Otherwise, use a numerical procedure for inversion

Figure 9   Schematic procedure to generate non-uniform random deviates from uniform random numbers.

Further, if the components of S are given by

$$S = \begin{bmatrix} \sigma_{11}^2 & \sigma_{12} \\ \sigma_{21} & \sigma_{22}^2 \end{bmatrix} \qquad (B11)$$

then the components of L are $\sigma_{11} = \sigma_{11}, a_{21} = \sigma_{12}/\sigma_{11}, a_{22} = \left[\sigma_{22}^2 - \sigma_{12}^2/\sigma_{11}^2\right]^{1/2}$
If N is matrix of standard normal deviates of mean 0 and unit variance, then the bivariate Gaussian deviates are generated by

$$R = N \cdot L^T \qquad (B12)$$

Suppose n_{11}, n_{12} are a pair of normal random numbers $N(0,1)$, then.lsl

$$b_1 = n_{11}n_{11} \qquad (B13)$$

$$b_2 = n_{11}a_{21} + n_{12}a_{22} \qquad (B14)$$

b_1, b_2 are bivariate Gaussian with a mean of zero and covariance given by the matrix S. It is obvious that the mean of (b_1, b_2) is zero. The variances of b_1 and b_2 can be shown to be σ_{22}^2 and σ_{11}^2 and their covariance to be $\sigma_{12} = \sigma_{21}$.

Author Index

247

250

Subject Index